Science and Technology Concepts–Secondary™

Investigating
Biodiversity and
Interdependence

Student Guide

National Science Resources Center

The National Science Resources Center (NSRC) is operated by the Smithsonian Institution to improve the teaching of science in the nation's schools. The NSRC disseminates information about exemplary teaching resources, develops curriculum materials, and conducts outreach programs of leadership development and technical assistance to help school districts implement inquiry-centered science programs.

Smithsonian Institution

The Smithsonian Institution was created by an act of Congress in 1846 "for the increase and diffusion of knowledge..." This independent federal establishment is the world's largest museum complex and is responsible for public and scholarly activities, exhibitions, and research projects nationwide and overseas. Among the objectives of the Smithsonian is the application of its unique resources to enhance elementary and secondary education.

STC Program™ Project Sponsors

National Science Foundation

Bristol-Meyers Squibb Foundation

Dow Chemical Company

DuPont Company

Hewlett-Packard Company

The Robert Wood Johnson Foundation

Carolina Biological Supply Company

Science and Technology Concepts–Secondary™

Investigating
Biodiversity and
Interdependence

Student Guide

The STC *Program*™

Smithsonian Institution
National Science Resources Center

www.carolinacurriculum.com

Published by Carolina Biological Supply Company
Burlington, North Carolina

NOTICE This material is based upon work supported by the National Science Foundation under Grant No. ESI-9618091. Any opinions, findings, and conclusions or recommendations expressed in this material are those of the authors and do not necessarily reflect views of the National Science Foundation or the Smithsonian Institution.

This project was supported, in part, by the **National Science Foundation.** Opinions expressed are those of the authors and not necessarily those of the foundation.

ISBN 978-1-4350-0641-6

Published by Carolina Biological Supply Company, 2700 York Road, Burlington, NC 27215. Call toll free 1-800-334-5551.

1206

Science and Technology Concepts—Secondary™
Investigating Biodiversity and Interdependence

The following revision was based on the STC/MS™ module *Organisms—Macro to Micro.*

Developer
Wordwise Inc.
(Susan Feldkamp/Curriculum Developer,
Karen Gotimer/CEO)

Scientific Reviewer
Carl T. Kloock
Assistant Professor of Biology
California State University, Bakersfield

Illustrator
Max-Karl Winkler

Writers/Editors
Amy Charles
Devin Reese
Linda Harteker

Photo Research
Jane Martin
Gina McNeely

National Science Resources Center Staff

Executive Director
Thomas Emrick

Program Specialist/Revision Manager
Elizabeth Klemick

Contractor, Curriculum Research and Development
Devin Reese

Publications Graphics Specialist
Heidi M. Kupke

Carolina Biological Supply Company Staff

Director of Product and Development
Cindy Morgan

Marketing Manager, STC–Secondary™
Jeff Frates

Curriculum Editors
Lauren Goldsmith
Gary Metheny

Managing Editor, Curriculum Materials
Cindy Vines Bright

Publications Designers
Trey Foster
Charles Thacker
Greg Willette

Science and Technology Concepts for Middle Schools™
Organisms—Macro to Micro
Original Publication

Module Development Staff

Developer/Writer
Henry Milne

Science Advisor
Robert Matthews
Department of Entomology
University of Georgia

Contributing Writers
Catherine Stephens
Daniel Doersch

Illustrator
Taina Litwak

STC/MS™ Project Staff

Principal Investigator
Sally Goetz Shuler, Executive Director, NSRC

Project Director
Kitty Lou Smith

Curriculum Developers
David Marsland
Henry Milne
Carol O'Donnell
Dane J. Toler

Illustration Coordinator
Max-Karl Winkler

Photo Editor
Christine Hauser

Graphic Designer
Heidi M. Kupke

STC/MS™ Project Advisors

George Andrykovitch, Associate Professor of Biology, George Mason University

D. Wayne Coats, Zoologist, Environmental Research Center, Smithsonian Institution

Daniel Doersch, Teacher, Green Bay Area Public School District, Wisconsin

Patricia Gossel, Curator, National Museum of American History

LeLeng Isaacs, Professor, Department of Biological Sciences, Goucher College

Jennifer Kuzuma, Staff Officer, The National Academies, Board on Biology

Craig Laufer, Professor, Department of Biology, Hood College

Lynn Lewis, Associate Professor, Mary Washington College

Robert Matthews, Professor, Department of Entomology, University of Georgia

Wanda Rinker, Teacher, West Windsor/Plainsboro School District, New Jersey

Patricia Shields, Course Coordinator, Introductory Biology, George Mason University

Eileen Smith, Teacher, Montgomery County, Maryland, Public Schools

Kay Stieger, Teacher, Montgomery County, Maryland, Public Schools

Paul Williams, Emeritus Professor, University of Wisconsin—Madison, Wisconsin Fast Plants® Program

Acknowledgments

The National Science Resources Center thanks the following individuals who contributed to the development of *Organisms—From Macro to Micro:*

John W. Cross
Author of popular educational Web sites
hosted by the Missouri Botanical Garden

Charles Drews
Professor of Invertebrate Zoology,
Neurobiology, and Bioethics
Iowa State University, Ames, Iowa

Dan Lauffer, Program Manager
Wisconsin Fast Plants™ Program
University of Wisconsin—Madison

Coe Williams, Program Coordinator
Wisconsin Fast Plants® Program
University of Wisconsin—Madison

Patricia A. Hagan, Senior Associate
The McKenzie Group, Washington, D.C.

Jonathan Jones, Principal (retired)
Cabin John Middle School, Potomac, Maryland

Robert Domergue, Principal
Robert Frost Middle School, Rockville,
Maryland

The National Science Resources Center gratefully acknowledges
the following individuals and school systems for their assistance
with the national field-testing of *Organisms—From Macro to Micro:*

The Einstein Project
Green Bay, Wisconsin

Site Coordinator
Sue Theno, Director
The Einstein Project

Green Bay Area Public School District
Daniel Doersch, Teacher
Seymour Middle School

Cindy Wallendal, Teacher
Lombardi Middle School

Mary Conard, Teacher
DePere Middle School

Montgomery County Public Schools
Montgomery County, Maryland
Eileen Smith, Teacher
Robert Frost Middle School

Eugene Public School District
Eugene, Oregon

Site Coordinator
Bob Curtis, Science Specialist
Lane Educational Service District
Eugene, Oregon

Angie Ruzicka, Teacher
Cal Young Middle School

Scott Baker, Teacher
Cal Young Middle School

Courtney Abbott, Teacher
Kelly Middle School

West Windsor-Plainsboro Regional School District
West Windsor/Plainsboro, New Jersey

Site Coordinator
Miriam A. Robin, Supervisor of Science
Grades 6-8
Community Middle School

Kevin MacKenzie, Teacher
Community Middle School

Shereen Rochford, Teacher
Grover Middle School

Wanda Rinker, Teacher
Community Middle School

Huntsville School District
Huntsville, Alabama

Site Coordinator
Sandy Enger, University of Alabama
Hunstville, Alabama

Rhonda Hudson, Teacher
Eva School

Jennifer Elam, Teacher
Meridianville Middle School

Cheryl Adams, Teacher
Liberty Middle School

North Carolina, Alamance-Burlington
School System

Mary Young, Teacher
Western Middle School

The NSRC appreciates the contribution of its
STC/MS project evaluation consultants—

Program Evaluation Research Group (PERG), Lesley College

Sabra Lee
Researcher, PERG

Center for the Study of Testing, Evaluation,
and Education Policy (CSTEEP), Boston College

Joseph Pedulla
Director, CSTEEP

Preface

Community leaders and state and local school officials across the country are recognizing the need to implement science education programs consistent with the National Science Education Standards to attain the important national goal of scientific literacy for all students in the 21st century. The Standards present a bold vision of science education. They identify what students at various levels should know and be able to do. They also emphasize the importance of transforming the science curriculum to enable students to engage actively in scientific inquiry as a way to develop conceptual understanding as well as problem-solving skills.

The development of effective standards-based, inquiry-centered curriculum materials is a key step in achieving scientific literacy. The National Science Resources Center (NSRC) has responded to this challenge through Science and Technology Concepts–Secondary™. Prior to the development of these materials, there were very few science curriculum resources for secondary students that embodied scientific inquiry and hands-on learning. With the publication of STC–Secondary™, schools will have a rich set of curriculum resources to fill this need.

Since its founding in 1985, the NSRC has made many significant contributions to the goal of achieving scientific literacy for all students. In addition to developing Science and Technology Concepts–Elementary™—an inquiry-centered science curriculum for grades K through 6—the NSRC has been active in disseminating information on science teaching resources, preparing school district leaders to spearhead science education reform, and providing technical assistance to school districts. These programs have had a significant impact on science education throughout the country. The transformation of science education is a challenging task that will continue to require the kind of strategic thinking and insistence on excellence that the NSRC has demonstrated in all of its curriculum development and outreach programs. The Smithsonian Institution, our sponsoring organization, takes great pride in the publication of this exciting new science program for secondary students.

Letter to the Students

Smithsonian Institution
National Science Resources Center

Dear Student,

The National Science Resources Center's (NSRC) mission is to improve the learning and teaching of science for K-12 students. As an organization of the Smithsonian Institution, the NSRC is dedicated to the establishment of effective science programs for all students. To contribute to that goal, the NSRC has developed and published two comprehensive, research-based science curriculum programs: Science and Technology Concepts-Elementary™ and Science and Technology Concepts-Secondary™.

By using the STC-Secondary™ curriculum materials, we know that you will build an understanding of important concepts in life, earth, and physical sciences; learn critical-thinking skills; and develop positive attitudes toward science and technology. The National Science Education Standards state that all secondary students "...should be provided opportunities to engage in full and partial inquiries.... With an appropriate curriculum and adequate instruction, ... students can develop the skills of investigation and the understanding that scientific inquiry is guided by knowledge, observations, ideas, and questions."

STC-Secondary also addresses the national technology standards published by the International Technology Education Association. Informed by research and guided by standards, the design of the STC-Secondary units addresses four critical goals:

• Use of effective student and teacher assessment strategies to improve learning and teaching

• Integration of literacy into the learning of science by giving students the lens of language to focus and clarify their thinking and activities

• Enhanced learning using new technologies to help students visualize processes and relationships that are normally invisible or difficult to understand

• Incorporation of strategies to actively engage parents to support the learning process

We hope that by using the STC-Secondary curriculum you will expand your interest, curiosity, and understanding about the world around you. We welcome comments from students and teachers about their experiences with the STC-Secondary program materials.

Thomas Emrick
Executive Director
National Science Resources Center

Navigating an STC–Secondary™ Student Guide

INTRODUCTION
This short paragraph helps get you interested about the upcoming inquiries.

MATERIALS
This helps you get organized and prepare for your inquiries.

READING SELECTION: BUILDING YOUR UNDERSTANDING
These reading selections are part of the lesson, and give you information about the topic or concept you are exploring.

NOTEBOOK ICON
During the course of an inquiry, you'll record data in different ways. This icon lets you know to record in your science notebook. Student sheets are called out when you're to write there. You may go back and forth between your notebook and a student sheet. Watch carefully for the icon throughout the procedure.

SAFETY TIPS
Safety in the science classroom is very important. Tips throughout the student guide will help you to practice safe techniques while conducting investigations. It is very important to read and follow all safety tips.

SAFETY TIP

PROCEDURE

This tells you what to do. Sometimes the steps are very specific, and sometimes they guide you to come up with your own investigation and ways to record data.

REFLECTING ON WHAT YOU'VE DONE

These questions help you think about what you've learned during the lesson's inquiries, apply them to different situations, and generate new questions. Often you'll discuss your ideas with the class.

READING SELECTION: EXTENDING YOUR KNOWLEDGE

These reading selections come after the lesson, and show new ways that the topic or concept you learned about during the lesson can be applied, often in real-world situations.

GLOSSARY

Here you can find scientific terms defined.

INDEX

Locate specific information within the student guide using the index.

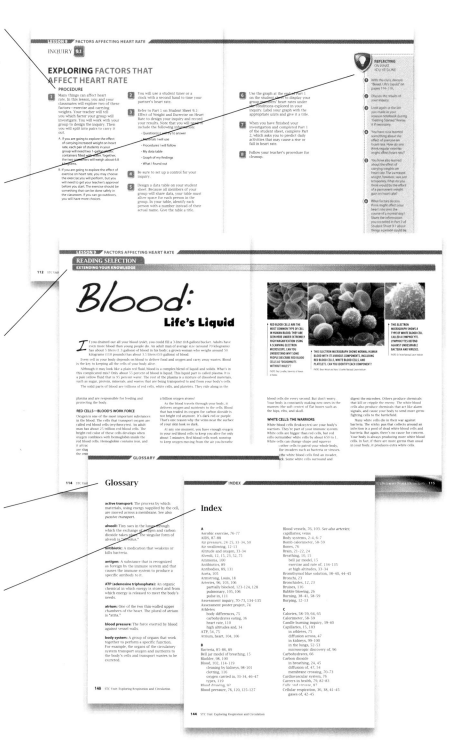

Contents

CONTENTS

WHAT ARE ORGANISMS?

▶ WHICH ORGANISMS
CAN YOU IDENTIFY IN
THIS ILLUSTRATION?

INTRODUCTION

What are organisms?
What makes them alive?
What do they need in their
environment to survive? How
do they reproduce and pass
along their characteristics
to their offspring? How are
they named and grouped
by scientists for easy
identification? These are
just a few of many questions
you will explore during this
unit. In this lesson, you will
observe photos of several
organisms. You'll examine
many of these organisms
in living detail later in the
unit. As you complete this
introductory activity, think of
questions about organisms
that you would like to answer
during this unit.

OBJECTIVES FOR THIS LESSON

▶ Develop a list of traits common to all living things.

▶ Construct a working definition of the word "organism."

▶ List some of the physical characteristics of the organisms shown on the organism photo cards.

▶ Assign each organism a genus and species name.

▶ Determine the appropriate place for each organism on the class habitat poster.

▶ MATERIALS FOR LESSON 1

For your group

1   set of organism photo cards
1   copy of Student Sheet 1.1: Latin and Greek Terms
1   loose-leaf ring
1   tag

GETTING STARTED

1 With your group, look over the organisms on the photo cards. Discuss the traits that these and all other living things have in common. Select at least five traits and list them in your science notebook. ✏

2 With your group, develop and record a working definition of the word "organism." A working definition is one that changes as you discover more information.

3 Discuss your ideas with the class.

▶ **STUDENTS WORKING WITH ORGANISM PHOTO CARDS**

PHOTO: Eric Long, Smithsonian Institution

DESCRIBING AND NAMING ORGANISMS

PROCEDURE

1. Follow your teacher's directions for identifying the physical characteristics of each organism pictured on the organism photo cards.

2. With your class, read "What's in an Organism's Name?" on pages 10-15.

3. Use Student Sheet 1.1: Latin and Greek Terms to assign each organism a logical scientific name, including both a genus and a species name. Print each name in the space provided on the inside of the organism card. Do not use both Latin and Greek words in the same name. You will learn the actual scientific names of some of these organisms as you work with them in greater detail in the unit.

4. Work with the class to agree on the most suitable habitat for each organism, as seen in Figure 1.1. Be prepared to explain your choices.

▶ THE CLASS AGREES THAT THIS ORGANISM WOULD PROBABLY LIVE IN AN AQUATIC HABITAT.
FIGURE **1.1**

Inquiry 1.1 continued

5 Print your class period and your group members' names on the identification tag. Open the loose-leaf ring. Insert the tag and the photo cards onto the ring and close it securely, as shown in Figure 1.2.

6 Return the tagged set of cards to your teacher for storage. Your teacher will give them back to you periodically to be revised.

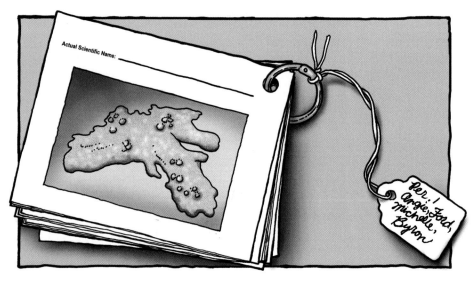

▶ **KEEP THE TAG FACING FRONT FOR EASY IDENTIFICATION.**
FIGURE **1.2**

REFLECTING
ON WHAT
YOU'VE DONE

1 Based on what you have learned in this lesson, work with your group to revise your list of traits common to all organisms and your working definition of organism as necessary. Record your revisions in your science notebook.

2 With the class, discuss the differences between a cougar, mountain lion, and puma, based on your homework research.

3 Based on what you've learned in this lesson, answer the following questions in your science notebook:

A. Why do scientists use Latin and Greek terms to create names for newly discovered organisms?

B. What do organisms need in their environment to be able to carry on their life processes and survive?

4 In your science notebook, list three or more questions about organisms that you would like to answer during this unit. Be prepared to share one question with the class.

5 Read "Taxonomy Taken a Step Further" on pages 16-19 and discuss your answers to the questions that follow with the class.

That's Life!

In this lesson, you learned that an organism is a "living thing." That sounds simple enough—until you ask yourself what "living" means.

What makes living things different from nonliving things? All living things grow, reproduce, and eventually die. They all respond in some way to changes in their environment.

They all have highly organized systems for growing, reproducing, and responding to the environment. So growth, reproduction, and response are referred to as "life processes."

Living things need energy to make these life processes go. All living things take in nutrients from their surroundings. Plants use energy from

▶ LIVING OR NONLIVING?

READING SELECTION
EXTENDING YOUR KNOWLEDGE

the sun to manufacture their own food, in the form of glucose (a kind of sugar), and use that glucose as an energy source. Other organisms, such as fungi and animals, take in food made by others, and break it down into small molecules during the process called digestion. Living things further break down some of these molecules—like glucose and bits of fat—to harvest the energy from them. And all living things, through these energy-getting activities, produce wastes that they must excrete, or get rid of, along the way. So food-getting, digestion, harvesting food's chemical energy, and excretion support the life processes common to living things.

There are many nonliving things that seem like they might fit the definition of life. A thermostat, for instance, has highly organized systems aimed at responding to changes in the environment, and it consumes electricity to get its power. You might say it excretes heat. But it doesn't reproduce, and has no way to reproduce; it isn't alive.

What about crystals? They're highly organized arrays of atoms. A crystal of salt has a lattice of chloride and sodium atoms that extends its pattern from one end of the crystal to another. You can even grow crystals; if you put one in a solution of its atoms, it will add the atoms to its pattern, and maintain that pattern. Eventually you'll see the crystal getting bigger and bigger. But the crystal does not reproduce; all it can do is add to itself. It doesn't take in energy to fuel its processes, and it doesn't excrete.

What about a robot? Imagine a robot-building robot. It takes in energy and uses it to power its motion. It excretes heat. Given the right

materials, it can create more robots just like itself. With built-in sensors, it can respond to heat, cold, and other stimuli. Given a relatively small set of rules—which may turn out not to be so different from the rules that guide insects or amoebas—robots can act independently, make decisions, and get around on their own. That's why we sent one to Mars as our first explorer on the planet's surface.

This is where we change the rules a little bit, and say that living organisms must also be composed of cells. It may seem a little discriminatory against robots, but that's biology for you.

A living thing may be composed of only one cell, like an amoeba, or millions of cells, like an elephant! But both the single-celled amoeba and the multi-cellular elephant perform all of the same life processes. Of course, they don't perform them all in exactly the same way.

▶ THIS ROBOT IS ABLE TO PERFORM SOME OF THE SAME TASKS AS HUMANS. WHAT MAKES THIS ROBOT DIFFERENT FROM A HUMAN BEING?

The organisms you will investigate in this unit will also vary in size, shape, and complexity. They all have something in common—the ability to carry out life processes to stay alive and healthy enough to produce more of their own species.

Now that you know what makes something living, it becomes obvious why some things are considered to be nonliving. Nonliving things have no cells, and cannot respond to the environment, or grow and reproduce on their own. ■

DISCUSSION QUESTIONS

1. Astronauts to the moon bring back a round, sticky thing with three protrusions that look like legs. How could they determine whether it is a living thing?

2. How else could you define the difference between living and nonliving things? Why did scientists choose the definition they did?

▶ DESPITE THEIR GREAT DIFFERENCE IN SIZE, THE ELEPHANT AND THE AMOEBA HAVE MUCH IN COMMON.

READING SELECTION
EXTENDING YOUR KNOWLEDGE

WHAT'S in an Organism's Name?

Even if you didn't know much about biology, you probably could guess that lions and tigers are close relatives. You also could be pretty sure that bears are animals and that roses are plants. It's easy because these organisms are familiar to us and have such distinctive appearances.

What if you had to find out whether mice, elephants, and bats were related? To answer this question, you would turn to taxonomy. Taxonomy is the science of classifying living things. Taxonomy is based on the principle that everything in our world is related in some way. It is a science that groups organisms according to their structures and functions.

Taxonomy was introduced in the 18th century by Carolus Linnaeus (1707-1778), a Swedish scientist. Linnaeus's interest in taxonomy started early. His father had a large garden, and he introduced his son to the science of plants. Linnaeus enjoyed studying plants, but even as a boy, he recognized that little information was available about how to classify plants. He saw a need for a universal classification system that would allow all scientists to communicate with one another about living things in a meaningful way.

▶ CAROLUS LINNAEUS
PHOTO: Courtesy of the National Library of Medicine

▶ THESE ANIMALS LOOK VERY DIFFERENT, BUT THEY'VE GOT ENOUGH IN COMMON TO BE IN THE SAME BIOLOGICAL CLASS. CAN YOU IDENTIFY THEIR COMMON FEATURES?

PHOTO: Charlotte Raymond, Photographer
PHOTO (top right): Justin B. Boyles
PHOTO (bottom right): Mark Bray/creativecommons.org

READING SELECTION
EXTENDING YOUR KNOWLEDGE

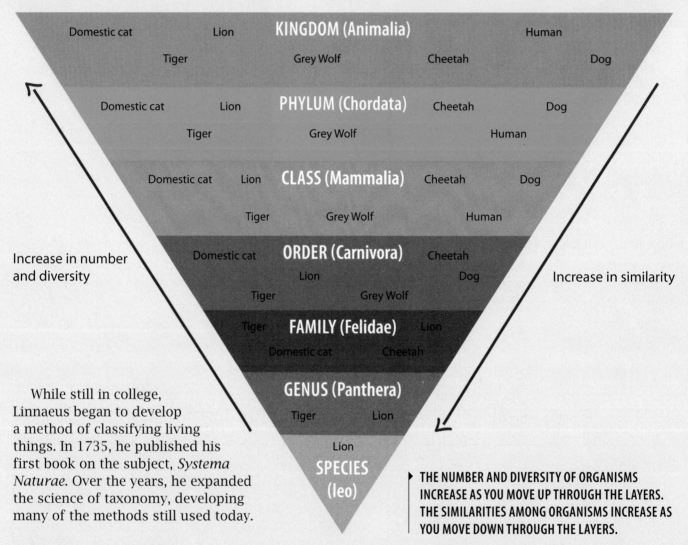

KINGDOM (Animalia)
Domestic cat   Lion   Human
    Tiger   Grey Wolf   Cheetah   Dog

PHYLUM (Chordata)
Domestic cat   Lion   Cheetah   Dog
    Tiger   Grey Wolf   Human

CLASS (Mammalia)
Domestic cat   Lion   Cheetah   Dog
    Tiger   Grey Wolf   Human

ORDER (Carnivora)
Domestic cat   Cheetah
        Lion   Dog
    Tiger   Grey Wolf

FAMILY (Felidae)
Tiger   Lion
Domestic cat   Cheetah

GENUS (Panthera)
    Tiger   Lion

Lion
SPECIES (leo)

Increase in number and diversity

Increase in similarity

THE NUMBER AND DIVERSITY OF ORGANISMS INCREASE AS YOU MOVE UP THROUGH THE LAYERS. THE SIMILARITIES AMONG ORGANISMS INCREASE AS YOU MOVE DOWN THROUGH THE LAYERS.

While still in college, Linnaeus began to develop a method of classifying living things. In 1735, he published his first book on the subject, *Systema Naturae*. Over the years, he expanded the science of taxonomy, developing many of the methods still used today.

A SEVEN-LAYER SYSTEM

Linnaeus developed rules for classifying plants and animals according to their structures. His work resulted in a seven-layer system: kingdom, phylum, class, order, family, genus, and species.

You can think of the system as an upside-down triangle. The top layer of the triangle is the kingdom category. Each kingdom contains the greatest number and diversity of organisms of the entire system. Because this layer is the largest, the organisms in it have fewer features in common than do organisms in the six other layers. For example, creatures as different as jellyfish and lions are both part of the Animal kingdom. As you move down the triangle, fewer organisms are included in each category, but the organisms within each category have more features in common.

Look at Table 1.1. You will see that all of the organisms listed across the top of the table are in the same kingdom, phylum, and class. With the exception of humans, they are also in the same order. The dog and grey wolf are in a different family than the four cats. And, of the four cats, only the tiger and the lion are in the same genus. The lion has the species name *leo*, which makes it unique from all the other animals in the chart.

Using Linnaeus's system, ants and spiders are part of the Animal kingdom. They are also both members of the phylum Arthropoda because they have jointed legs. But each is in a different class. Ants are in the class of animals with three-part bodies and six legs. This class is called Insecta. Spiders are in the class of eight-legged organisms with two-part bodies. This class is known as Arachnida.

ADDING NEW KINGDOMS

Linnaeus grouped all organisms into two main kingdoms—Plants and Animals. Until the second half of the 20th century, most biologists used his system. Then they added a third kingdom, the Protists, because microorganisms did not all clearly fit into the Animal or Plant kingdoms. As scientists discovered more and more information about organisms, they added two more kingdoms—Fungi and Monerans. Since then, the number of kingdoms and the organisms that should be included in each have been under debate. Scientists now propose between six and a few dozen kingdoms. While there is not consensus at the kingdom level, scientists do agree that organisms can be placed into one of three domains—Bacteria, Archaea, or Eukarya—which are broader than kingdoms.

TABLE 1.1 THE SEVEN-LAYER SYSTEM

	DOMESTIC CAT	LION	CHEETAH	TIGER	DOG	GREY WOLF	HUMAN
KINGDOM	Animalia	Animalia	Animalia	Animalia	Animalia	Animalia	Animalia
PHYLUM	Chordata	Chordata	Chordata	Chordata	Chordata	Chordata	Chordata
CLASS	Mammalia	Mammalia	Mammalia	Mammalia	Mammalia	Mammalia	Mammalia
ORDER	Carnivora	Carnivora	Carnivora	Carnivora	Carnivora	Carnivora	Primates
FAMILY	Felidae	Felidae	Felidae	Felidae	Canidae	Canidae	Hominidae
GENUS	*Felis*	*Panthera*	*Acinonyx*	*Panthera*	*Canis*	*Canis*	*Homo*
SPECIES	*silvestris*	*leo*	*jubatus*	*tigris*	*familiaris*	*lupus*	*sapiens*

READING SELECTION

EXTENDING YOUR KNOWLEDGE

▶ THE TWO-PART BODY AND EIGHT LEGS IDENTIFY THIS SPIDER AS BELONGING TO PHYLUM ARTHROPODA, CLASS ARACHNIDA.

PHOTO: Tijl Vercaemer/
creativecommons.org

▶ THEIR THREE-PART BODY AND SIX LEGS PUT ANTS IN PHYLUM ARTHROPODA AND CLASS INSECTA. THE TWO LARGE SECTIONS AT EACH END ARE THE HEAD AND ABDOMEN. THE SMALLER SEGMENTS IN BETWEEN COMPRISE A THIRD PART, THE THORAX.

PHOTO: B.M. Drees, Texas AgriLife Extension Service, Texas A & M University

ONE ORGANISM, TWO NAMES

Linnaeus also developed a system for naming organisms, under which a two-part scientific name is assigned to every organism. An organism is named on the basis of its genus and species. The species name is usually an adjective, and the genus name is usually a noun. The first letter of the genus name is always capitalized. When the scientific name is typed, it is always in italics. When written by hand, it is underlined.

Some organisms are named after one of their prominent features. The scientific name for the red maple, for example, is *Acer rubrum. Acer* means "maple," and *rubrum* means "red." Some are named after the location in which they are found. A species of fly discovered in Humbug Creek, California, was named *Oligodranes humbug.* Others are named after the scientist who discovered them.

There are other sources of names, too. There is a spider, *Draculoides bramstokeri*, named after the novel *Dracula* by Bram Stoker. Perhaps the grandfather of all names belongs to an aphid, a tiny insect. Its scientific name is *Myzocallis kahawaluokalani.* This Hawaiian name supposedly means, "You fish on your side of the lagoon and I'll fish on the other, and no one will fish in the middle."

Linnaeus's groundbreaking work of the 18th century remains the basis of the system we use today. Taxonomy now helps scientists classify more than 10 million species of organisms on Earth, and new kinds are discovered every year. ∎

DISCUSSION QUESTIONS

1. Why did Linnaeus think it was useful to classify organisms?

2. How can taxonomy help us understand the relationships among organisms?

Taxonomy
TAKEN A STEP FURTHER

A scientist who studied flies and mosquitoes revolutionized our understanding of taxonomy. Willi Hennig (1913–1976), a German entomologist (someone who studies insects), spent the end of World War II as a prisoner of war. While a prisoner, he painstakingly drafted the 170-page document that would change the future of taxonomy. Up until Hennig's time, organisms were classified based on the Linnean system of shared characteristics with which you are familiar. Hennig proposed that organisms be further classified according to shared ancestry. He founded what is now known as "cladistics," or the science of evolutionary trees.

An evolutionary tree, or cladogram, proposes how groups of organisms are related based on certain characteristics. A cladogram has a built-in time axis. Time is advancing from the base of the tree to the branches. For example, the cladogram below proposes relationships between the Tuatara lizard, some other now-extinct animals, and mammals (represented by the possum).

The taxonomy in this cladogram is based on scientists' study of the joint between the skull and jawbone. It is thought that certain jawbones (articular and quadrate) evolved into the tiny bones that vibrate in the ears of mammals in response to sound. Starting with the most ancient (Sphenodon), and moving up through the tree, you can see how the bones (shown in red and blue) got smaller and eventually became part of the ear. These changes happened over long periods of time, or "evolutionary time," through a process of adaptation.

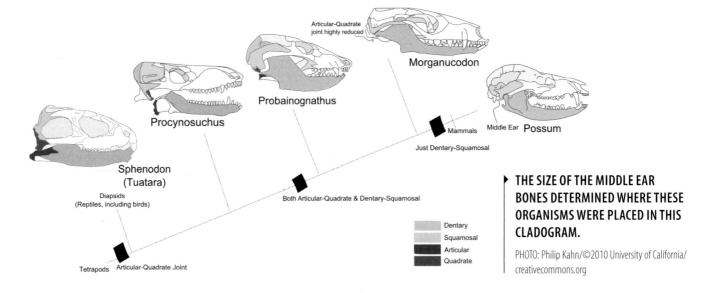

Articular-Quadrate joint highly reduced

Morganucodon

Probainognathus

Procynosuchus

Mammals

Middle Ear Possum

Just Dentary-Squamosal

Sphenodon (Tuatara)

Diapsids (Reptiles, including birds)

Both Articular-Quadrate & Dentary-Squamosal

Tetrapods   Articular-Quadrate Joint

Dentary
Squamosal
Articular
Quadrate

▶ **THE SIZE OF THE MIDDLE EAR BONES DETERMINED WHERE THESE ORGANISMS WERE PLACED IN THIS CLADOGRAM.**

PHOTO: Philip Kahn/©2010 University of California/ creativecommons.org

▶ THIS FOSSILIZED FISH WAS DISCOVERED IN CHINA AND
IS BELIEVED TO HAVE LIVED 100 MILLION YEARS AGO.
WHAT CLUES MIGHT IT YIELD?

PHOTO: Bruce Avera Hunter/life.nbii.gov

How is this related to Linnean taxonomy? Think of Linnean taxonomy as a photograph, and cladistics as a movie. Linnean taxonomy gives us a snapshot of all organisms, but tells us nothing about how they came to be. Cladistics plays evolution's movie. It shows us the relationships between organisms from different kingdoms, classes, phyla, and so on. The Tuatara lizard may be a reptile, but with cladistics we can see how its descendants evolved into a mammal.

Cladograms do not have to be based on bone size. Indeed, they can be based on any characteristics that show a pattern of change; this might be number of toes, length of legs, or placement of eyes on the head. Because cladograms show relationships that happened in the past, in many cases they include animals that are extinct. Scientists trying to build cladograms rely on present-day organisms and fossils. Fossilized skulls, bones, feathers, teeth, or even fur could help to build a cladogram.

Deciding which characteristics to focus on and figuring out how they differ among fossil organisms is like detective work. As additional clues (such as more fossils) become available, cladograms are modified. For example, in 2007, a fossil jawbone from a species called *Homo habilis* was discovered in Kenya, a country in Africa. In cladistics, *Homo habilis* was considered to have evolved into *Homo erectus*, who evolved into a species currently on the

READING SELECTION

EXTENDING YOUR KNOWLEDGE

earth (*Homo sapiens*)—us! But scientists were able to date the fossil bone, which suggested that *Homo habilis* lived at the same time as *Homo erectus* and was not its ancestor.

Fossils cannot provide all the information needed to build cladograms. Often fossils are absent or incomplete. In recent years, scientists have turned to genetic evidence to figure out how organisms are related. The 1953 discovery of the structure of DNA paved the way for a whole new method of understanding the relationships amongst groups of organisms.

DNA, or deoxyribonucleic acid, is a very long molecule in the cells of organisms that contains their genetic code. Every cell in our body needs to have a complete genetic code. DNA is like a chain made up of four kinds of links. The order in which the links occur forms the code. If we

▶ HOW THE DNA LINKS ARE ARRANGED IN YOUR 20,000+ GENES DETERMINES HOW TALL YOU ARE, THE COLOR OF YOUR EYES, AND MANY OTHER THINGS ABOUT YOU.

PHOTO: Steve Hillebrand/U.S. Fish and Wildlife Service

picture them as red, green, blue, and yellow links, a sequence of *red, red, green* would code a different meaning than a sequence of *green, yellow, blue.* Long sections of this code are called genes, and the sequence of links in each gene serves as a plan for making a type of protein. Proteins, working together, give rise to our particular characteristics, such as eye color and height.

When DNA's structure and function were understood, scientists saw right away that comparing the genes of different species could tell us something about how they were related, evolutionarily. The problem was that the research was slow and expensive. Then, in 1983, a scientist named Kary Mullis invented a technique called "polymerase chain reaction" (PCR, for short) that could very quickly make thousands of copies of short sections of DNA. Suddenly, scientists could easily run thousands of experiments comparing the DNA of one species with that of another.

How exactly does this help find evolutionary relationships? A species' DNA does change, but slowly. Species that are more closely related have more similar DNA. It just hasn't had time to change very much. The DNA of humans and chimpanzees, for example, is 96% the same. What makes us distinct from chimpanzees is mapped out in less than 4% of our DNA. In other words, humans have not evolved very far away from chimpanzees.

A modern cladogram may be constructed from genetic information, information on visible characteristics, or both types of information. The strongest approach is a combination of methods in which one confirms the other for a particular group of organisms. To build a cladogram, a scientist gathers a body of evidence about those organisms. Then she tries to find traits they share, and looks carefully to see how many traits each species shares with the others. The oldest

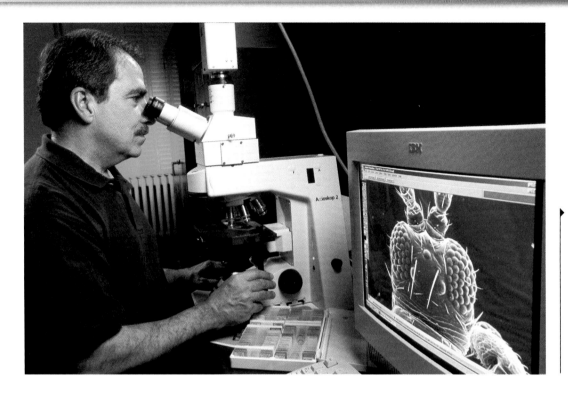

▶ SCIENTISTS CONTINUE TO STUDY CHARACTERISTICS OF ORGANISMS TO LEARN ABOUT THE RELATIONSHIPS AMONG THEM.

PHOTO: Peggy Greb, Agricultural Research Service/U.S. Department of Agriculture

common ancestor will be the starting point of a branch on the evolutionary tree, and will share only one trait with the rest of organisms: the other traits will have evolved along the way.

Polymerase chain reaction continues to tell us a lot about ancestral relationships among organisms. Biologists need as much information on the species they're studying as possible. These days, supercomputers allow for the processing of very large amounts of information (e.g., thousands of bone measurements) to help build cladograms. Remember that cladistics is detective work, and a cladogram is a hypothesis for how groups of organisms are related. More evidence may make for a better hypothesis. Cladograms are revised as more reliable or meaningful information is found through fossils, DNA analysis, or other techniques. How many years of study will it take to get cladograms just right? ■

DISCUSSION QUESTIONS

1. How has taxonomy changed since Linneaus started it?

2. Mushrooms that have inky fluid in their caps have long been classified in the same genus of "inkcaps." But the story told by DNA analysis is different; it proposes that they belong to several distinct evolutionary lines. Which story do you believe and why?

THE WOWBUG: GETTING A CLOSER LOOK

INTRODUCTION

Many of the organisms pictured on the organism photo cards in Lesson 1 cannot be seen very well, or at all, with the naked eye. Those photos were made with the aid of magnification. To view the organisms yourself, you would use a microscope. In this lesson, you will learn how to prepare a dry-mount slide and how to use a compound light microscope to observe organisms. You will also learn how to prepare scientific drawings according to a specific set of guidelines, which you will use throughout the unit. You will learn these skills while observing and learning about an interesting organism called the "WOWBug," a tiny wasp that is harmless to humans.

▸ **A COMPOUND LIGHT MICROSCOPE ESTIMATED TO BE 40–50 YEARS OLD**

PHOTO: Courtesy of Henry Milne/© NSRC

OBJECTIVES FOR THIS LESSON

- Learn the parts of a microscope, and practice manipulating them to obtain the best image of slide-mounted specimens.

- Measure the diameter of the field of view under different magnifications of the compound microscope.

- Learn how to handle, manipulate, and recapture WOWBugs.

- Prepare dry-mount slides of live WOWBugs.

- Observe WOWBug grooming behavior.

- Draw, label, and measure a WOWBug, following specific guidelines for scientific drawings.

- Update your organism photo card for WOWBugs.

MATERIALS FOR LESSON 2

For you

1	copy of Student Sheet 2.3a: Guidelines for Scientific Drawings
1	copy of Student Sheet 2.3b: Drawing Your WOWBug

For your group

2	compound light microscopes
2	depression slides
2	plastic slides
4	hand lenses
5	WOWBugs
2	pipe cleaners
1	sheet of notebook paper
2	pieces of transparent tape
2	transparent rulers
2	metric rulers, 30 cm (12 in.)
1	plastic cup with lid, 4 oz
1	plastic cup of flour
2	toothpicks
1	box of colored pencils
1	set of organism photo cards

GETTING STARTED

1 Working in your group, observe your hand lens. Notice the shape of each lens. They are called "convex lenses" because they bulge in the middle and taper toward the edge. Stand at your desk and hold your hand lens about 1 centimeter (cm) above a line of text in your Student Guide.

2 In your science notebook, make a quick sketch of the hand lens, showing the appearance of the text through each of its lenses. Record your estimates of the magnifications directly on the corresponding lenses in your sketch. For example, if you estimated that one of the lenses magnified the text to two times its normal size, you would write 2x on that part of your sketch. ✎

3 Center the smaller lens on top of the word "of" in this sentence. Close one eye, but continue to stare at the word "of." Slowly raise the lens from the page toward your open eye. Record your answer to the following question:

A. How does the image of the word "of" change as you raise the lens?

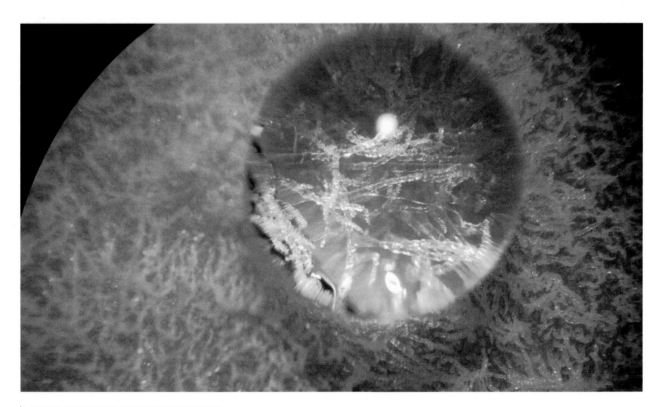

▶ **WHAT ABOUT THE SHAPE OF THE WATER DROPLET CAUSES IT TO MAGNIFY THE LEAF?**

PHOTO: Jodiepedia/creativecommons.org

4 Share hand lenses with your partner. Repeat Step 3, but stop raising the lens when the word "of" appears upside down and backward, while still remaining in focus. Pick up the second hand lens. Center its larger lens directly over the small lens of the first hand lens, which should still be focused on the word "of." Look through the large lens while you raise it slowly. Keep your head up. If you put your eye down to the lens you will not see the intended effect. Record your answer to the following question:

A. What happens to the image of the word "of" in the larger lens as you raise it away from the smaller lens?

5 With a partner, take a close look at your microscope. Refer to the reading selection "Through the Compound Eye" (on pages 24-25), which you read for homework, to identify the microscope's main parts and to find out how to calculate the magnifications you would get using its different lenses. Discuss with your group how two lenses work together in a microscope to produce an image.

READING SELECTION

BUILDING YOUR UNDERSTANDING

THROUGH THE COMPOUND EYE

For thousands of years, human beings have used tools. For a biologist, one of the most important tools is the microscope. Since its invention in the early 1600s, the microscope has been transformed into a relatively inexpensive, yet efficient, way for scientists such as yourself to view a world invisible to the naked eye.

You probably will use a compound light microscope during this unit. In this type of microscope, light is provided by either a mirror or a small, built-in lightbulb. The word "compound" refers to the two lenses—one in the eyepiece and one in an objective—that together magnify the image. You can calculate the total magnification by multiplying the magnification of the lens of the eyepiece by that of the lens in the objective.

The drawing on page 25 shows the parts of a compound microscope and explains the function of each part.

As you use your microscope during this unit, you will gain a working knowledge of its parts and their functions and become much more proficient at using this important tool of science. ■

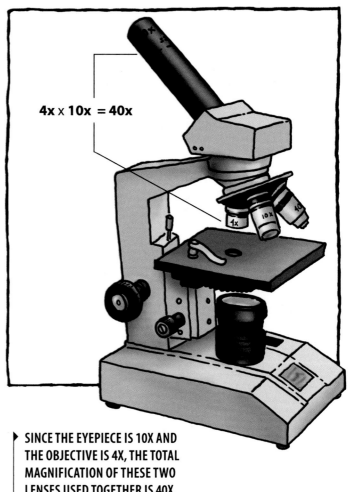

4x x 10x = 40x

▶ SINCE THE EYEPIECE IS 10X AND THE OBJECTIVE IS 4X, THE TOTAL MAGNIFICATION OF THESE TWO LENSES USED TOGETHER IS 40X.

The Compound Light Microscope

Eyepiece—Contains a 10x lens

Nosepiece—Holds the objective lenses; rotates to enable changing magnification

Objective Lenses—Used in combination with the eyepiece, they provide a range of magnifications, usually between 40x and 400x

Stage—Supports the slides

Diaphragm—Wheel or lever that adjusts amount of light that passes through hole in stage; provides proper contrast

Arm—Supports the upper part of the microscope; serves as a handle

Stage Clip—One on each side of hole in stage; helps to hold slides in place

Coarse Adjustment Knob—Raises and lowers stage or objective lenses

Light—Sends light through hole in stage to illuminate specimen on slide

Base—Supports the microscope; serves as a handle

Fine Adjustment Knob—Raises and lowers stage or objective lenses a tiny distance for exact focusing

INQUIRY **2.1**

CORRALLING YOUR WOWBUGS

PROCEDURE

1 Place a piece of notebook paper in front of your group. Your teacher will put about five female WOWBugs on the paper.

2 Have one member of the group very gently corral the WOWBugs into the center of the paper using the tip of the pipe cleaner, as seen in Figure 2.1. A gentle nudge of the pipe cleaner will stimulate the WOWBugs to change direction. After about 30 seconds, quickly pass the pipe cleaner to another group member. Have this member continue to corral the WOWBugs into the center of the paper. Continue until all group members have had a turn.

3 While the last group member is practicing handling the WOWBugs, have another group member carefully invert the plastic cup over each WOWBug, one at a time, until they have all crawled up on its inside surface. Save the WOWBugs for Inquiry 2.2.

▶ MOVE THE PIPE CLEANER QUICKLY TO KEEP UP WITH THE WOWBUGS!
FIGURE **2.1**

PREPARING A DRY-MOUNT SLIDE TO VIEW WOWBUG GROOMING BEHAVIOR

PROCEDURE

 Working with a partner, take the following steps to prepare a dry-mount slide of a WOWBug:

A. Place a depression slide on the notebook paper. Keep handy a flat, plastic slide and two 2-cm pieces of transparent tape.

B. Dip a toothpick into the container of flour supplied by your teacher. Then tap a few specks of the flour into the well of the depression slide. Stir the flour with the tip of the toothpick to scatter it around the depression. Use just enough flour dust so that the WOWBug will become slightly "dirty." Too much flour could harm the WOWBug.

C. Use the pipe cleaner to remove one WOWBug from the cup and transfer it into the slide's depression. Quickly place the flat, plastic slide on top of it, trapping the WOWBug in the depression.

D. Use the two pieces of transparent tape to fasten the ends of the slides together to prevent the WOWBug from escaping. Your slide should look like the one in Figure 2.2.

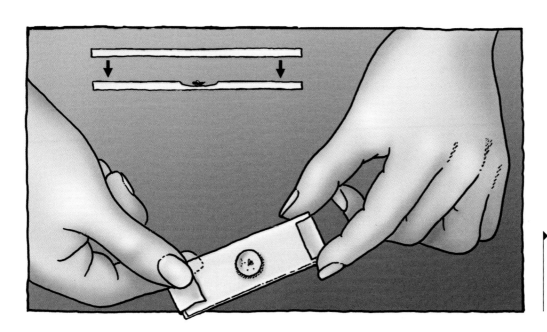

▶ FASTEN THE TAPE SECURELY OR THE WOWBUG MIGHT ESCAPE!
FIGURE **2.2**

Inquiry 2.2 continued

2 Take the following steps to view the dry-mount slide:

A. Place the dry-mount slide on the microscope stage and focus on the WOWBug under the lowest magnification. If the WOWBug is moving around, practice keeping it in the field of view by moving the slide slowly and smoothly with your fingers while you observe it through the eyepiece. If you are lucky, you may get to see the WOWBug standing still to clean the flour from its legs or antennae. Note whether it cleans away the specks of flour in any particular order or manner.

B. Repeat Procedure Step 2A with each of the other objective lenses until you have viewed the WOWBug under all magnifications.

C. Have your partner repeat Procedure Steps 2A and 2B.

3 Save the slide for Inquiry 2.3.

PREPARING SCIENTIFIC DRAWINGS OF THE WOWBUG

PROCEDURE

1 Listen while your teacher reviews the information on Student Sheet 2.3a: Guidelines for Scientific Drawings.

2 Have one student in your pair place your WOWBug dry-mount slide on the microscope stage.

3 Draw the WOWBug in the top circle on Student Sheet 2.3b: Drawing Your WOWBug, following the guidelines on Student Sheet 2.3a. Include as much detail as you can. Take turns with your partner at the microscope. The WOWBug should stand relatively still while cleaning itself, so you can pay close attention to details like the number of parts in the antennae or legs. Use the highest magnification through which you can see the entire WOWBug in greatest detail. This should be the one in which the WOWBug nearly fills the field of view. Since the WOWBug is three-dimensional, you may find that you have to adjust the fine focus at times to see the various structures more clearly.

4 If the WOWBug is too active and repeatedly crawls out of the field of view, try viewing the slide prepared by the other pair in your group. If you still can't see the WOWBug, ask your teacher for assistance.

5 Normally, as you view the WOWBug through the microscope, it will be right-side up so that you would be looking at its back, which scientists call its dorsum. If the WOWBug is walking upside down on the top cover, you will see its underside, which is called its venter. If this happens, flip the slide over on the stage and observe it from the other side. Title your drawing, "WOWBug—Dorsal View."

6 Label at least five structures. Refer to the illustration of the WOWBug in the reading selection "Dr. Matthews and the WOWBug" on pages 35–36 for names of its structures.

Inquiry 2.3 continued

 7 To complete your drawing, it is necessary to give the viewer or reader some idea of the size of the WOWBug. After following the steps below, you will better understand the relative sizes of different kinds of organisms when given a scientific drawing of them.

A. Remove the slide from the microscope stage, and switch the magnification to the lowest power.

B. Center the transparent ruler on the stage and count the number of millimeter (mm) lines you can see across the widest part of the field of view. Record this number on Student Sheet 2.3b on the line next to this label: Diameter of the Field of View (mm) at Low Magnification.

C. Switch the magnification to medium power, then repeat your measurement. Record this number next to this label: Diameter of the Field of View (mm) at Medium Magnification.

D. Switch the magnification to the highest power, then repeat this process once more. Record this number on your student sheet next to this label: Diameter of the Field of View (mm) at High Magnification (400x, 430x, or whichever is greatest on your microscope).

E. Remove the ruler. Switch the magnification to the lowest power, put the slide back on the stage, and focus on the WOWBug. Based on the number of ruler lines you counted under the lowest magnification, estimate the length of the WOWBug in mm. Record this number on your student sheet next to this label: Estimated WOWBug Body Length (mm) at Low Magnification.

F. Now place the ruler underneath the slide until the tip of one end of the WOWBug is right in the center of one of the mm marks, as shown in Figure 2.3.

G. Measure the actual body length of the WOWBug in mm, and record that length on your student sheet next to this label: Actual WOWBug Body Length (mm) at Low Magnification. Also record the length on Student Sheet 2.3b, just to the right of the drawing's title. Check the actual measurement against your estimate.

▶ YOU CAN ONLY MEASURE THE LENGTH OF THE WOWBUG ACCURATELY WHEN IT IS POSITIONED CORRECTLY AGAINST A MM MARK OF THE RULER.
FIGURE **2.3**

8 When you and your partner have completed your drawings, review the 10 guidelines on Student Sheet 2.3a. When you are satisfied that you have correctly followed these guidelines, trade drawings with your partner. Notice that the numbers 1–10 are listed below the title line of your drawing. These represent the drawing guidelines. Circle, in pencil, the number of each guideline that your partner did not follow. Then, assign each other's drawing a score of 1 to 10, based on the number of guidelines that were followed correctly. When you are both finished, return each other's drawings. Revise your drawing and give it back to your partner to review again. Erase the circle around each number of a guideline that was revised correctly. Continue until you both earn 10 points. This process is called peer evaluation. It is an important part of scientific inquiry.

9 Focus the microscope on one structure of the WOWBug (an antenna, a wing, or a leg, for example) using the highest magnification possible. Draw that structure in the second circle on Student Sheet 2.3b. Give it an appropriate title, and label anything on this structure that you found in the illustration of a WOWBug in the reading selection "Dr. Matthews and the WOWBug," which you read for homework. Peer-evaluate this drawing before turning it in, if you wish.

10 Update your organism photo card for the WOWBug.

REFLECTING
ON WHAT
YOU'VE DONE

1 Based on what you have learned in this lesson, answer the following questions in your science notebook. Be prepared to discuss your answers with the class.

A. Explain two ways in which the compound light microscope is an improvement over the microscope developed by Antony van Leeuwenhoek.

B. List three ways in which lenses are used as tools of science, in addition to their use in compound microscopes.

C. How did the diameter of the field of view change when you changed the microscope's objective lenses?

D. What characteristics of the WOWBug did you observe that suggest it is an insect?

E. In what ways did your WOWBug remove the flour dust from its body? List two reasons why you think grooming would be important to a WOWBug. (Hint: Why is grooming important for you?)

F. According to the reading selection "Intriguing Insects" on pages 32–34, how are parasitic wasps, such as WOWBugs, important to humans?

Intriguing Insects

When you think about insects, which come to mind first? Butterflies? Ants? Bees? In fact, beetles are the most common insect. If you lined up every kind of plant and animal in a row, every fourth organism would likely be a beetle. And beetles are only one kind of insect!

There are hundreds of types of insects on Earth, ranging from the common to the exotic. You're probably quite familiar with wasps, flies, mosquitoes, moths, crickets, fireflies, and dragonflies. Have you ever heard of a cicada known as the "buffalo head," whose head resembles a buffalo's, complete with a set of horns? Or the whirligig beetle, which uses its two sets of eyes in a clever way when it goes swimming? One set looks above the water's surface, while the other checks out the action below. And don't forget the fruit fly, *Drosophila*. The scientific study of the brief life cycle of this tiny fly laid the groundwork for modern genetics.

What do all of these insects have in common? They all have three distinct body parts—a head, a thorax, and an abdomen. They also have six legs, four wings, and an outer covering called an exoskeleton.

When you think about it, insects are just about everywhere. They live in our houses, in our gardens, on our pets, and sometimes even on us. You find them in lakes, ponds, and streams. They survive on the coldest mountains and in the hottest deserts.

▶ **THESE ARE JUST A FEW OF THE THOUSANDS OF VARIETIES OF BEETLES FOUND ALL OVER THE WORLD.**

PHOTO: Chip Clark, National Museum of Natural History, Smithsonian Institution

GOOD GUYS AND BAD GUYS

Some people don't like insects at all. However, each kind of insect has a role to play in the world, and each affects our lives in a different way.

Some insects seem to cause more than their share of trouble. According to Dr. Robert Matthews, a professor at the University of Georgia, insects have caused an enormous amount of human suffering. Some mosquitoes transmit diseases, like malaria and yellow fever, which are major threats to human health in much of the world. Flying grasshoppers called migratory locusts destroy entire fields of crops.

We consider other insects to be good guys. Honeybees pollinate the flowers of many of our favorite food crops. Anyone who has enjoyed a biscuit with honey also appreciates their efforts.

Less familiar insects, such as parasitic wasps, lay their eggs in or on other insects. A parasite is an organism that obtains its nutrients from another organism, generally damaging the other organism in the process.

▶ AS YOU CAN SEE, GRASSHOPPERS CAN DO CONSIDERABLE DAMAGE TO RANGELAND GRASSES SUCH AS WHEATGRASS, AS WELL AS CROPS.

PHOTO: Jack Dykinga, Agricultural Research Service/U.S. Department of Agriculture

▶ FEW CROPS CAN STAND UP TO A SWARM OF INSECTS SUCH AS THIS.

PHOTO: Michel Lecoq (CIRAD)

READING SELECTION

EXTENDING YOUR KNOWLEDGE

A WORLD WITHOUT WASPS

Parasites may sound destructive, but they also play an important role. For example, a world without parasitic wasps would be a very different place. These insects help lower Earth's pest population. In fact, scientists have calculated that a single pair of houseflies, if left alone, could potentially produce enough descendants in a year to cover the surface of the earth several centimeters deep. Fortunately, this doesn't happen, thanks to natural enemies such as parasitic wasps, which kill large numbers of flies every year. ■

DISCUSSION QUESTIONS

1. In many parts of the world, much effort is put into killing mosquitoes to control malaria. What unintended consequences might that have?

2. Insects don't have bones; their exoskeleton is the only skeleton they have. What reasons might there be for creatures like humans (vertebrates) to have developed skeletons inside our bodies, with soft outsides?

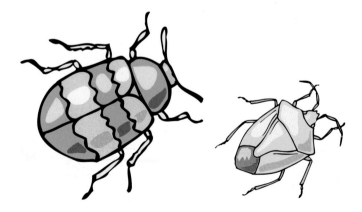

Dr. Matthews and the WOWBug

D r. Robert Matthews is an entomologist, a scientist who studies insects. He has studied insects for many years and in many parts of the world.

One of Dr. Matthews's favorite insects is a small parasitic wasp called *Melittobia digitata*. That's quite a mouthful, which is why Dr. Matthews nicknamed it the "WOWBug." He and his students have learned much about the strange habits of this intriguing insect. Through their efforts, the WOWBug has become one of the newest organisms studied in the science classroom.

What's so special about WOWBugs? And how did they make their way into the classroom? It was an unlikely beginning. Dr. Matthews did not find the bugs—they found him! While he was a graduate student, Dr. Matthews decided to examine the nests of some little wild bees he found outdoors. He took the nests inside and put them on a shelf in his laboratory. Later, he got the nests down to study them. To his surprise, he found not little bees, but WOWBugs! Unnoticed, they had sneaked into the nests, fed, and multiplied. They had destroyed nearly all of his bees, and Dr. Matthews was pretty angry.

▶ THESE WOWBUGS ARE ONLY 1.5 MILLIMETERS (0.06 INCHES) LONG, BUT THEY PLAY A VERY LARGE ROLE IN HELPING TO CONTROL BEE AND FLY POPULATIONS. FEMALE (LEFT); MALE (RIGHT). NOTE THE MALE'S UNUSUAL ANTENNAE.

PHOTO: Courtesy of Carolina Biological Supply Company

▶ DR. MATTHEWS, SECOND FROM LEFT, SHARING A BUTTERFLY COLLECTION WITH MEMBERS OF THE WOWBUGS TEAM.

PHOTO: Courtesy of R.W. Matthews

READING SELECTION

EXTENDING YOUR KNOWLEDGE

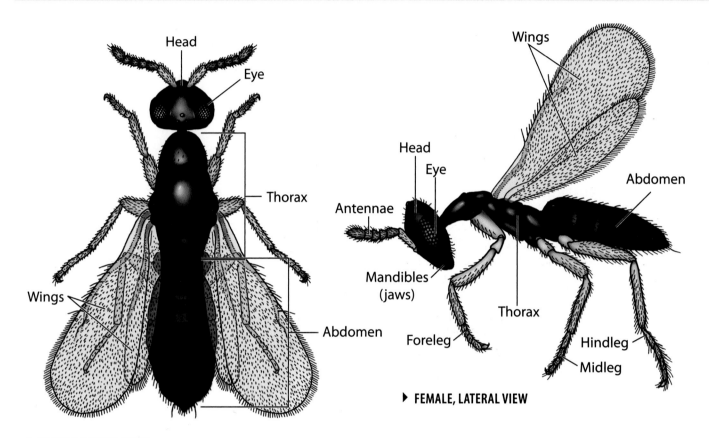

▶ **FEMALE, LATERAL VIEW**

▶ **FEMALE, DORSAL VIEW**

Many years later, while thinking about new ways to teach biology, Dr. Matthews remembered the WOWBug. He realized that the same WOWBug behaviors that nearly ruined his early research would make these little parasites wonderful in the science classroom. WOWBugs breed easily in large numbers, they have a very short life cycle, and they don't take up much space. Best of all, they can't hurt humans with their stingers.

As he worked with WOWBugs, Dr. Matthews continued to learn new and fascinating things about their biology and behavior. He wanted to share what he was learning. With the help of other scientists and teachers, Dr. Matthews developed a set of teaching activities to help students learn science concepts and skills by working with WOWBugs.

Scientists on the WOWBugs team at the University of Georgia continue to make new discoveries every day. They write a newsletter, give workshops for teachers, and develop new lab investigations. Wow! There is a lot to learn from such a tiny insect! ■

DISCUSSION QUESTIONS

1. What about WOWbugs has made them an excellent subject for laboratory research?

2. How might the WOWBugs' shape help them to be good parasites?

Microscope PIONEERS

You can't study organisms thoroughly without a good microscope. This tool, which today's scientists take for granted, has played a major role in helping scientists understand more about living things.

Robert Hooke and Antony van Leeuwenhoek were important pioneers in the development of this important scientific instrument. Hooke was born in England in 1635. A member of the Royal Society of England, he was one of the most famous scientists of his time. Leeuwenhoek was born in the Dutch town of Delft in 1632.

HOOKE: DISCOVERING THE MYSTERIES OF CORK

Today, Robert Hooke is remembered more as a mathematician than as a biologist. But like all scientists of his day, he had broad interests. He made many contributions to biology. In his book, *Micrographia*, Hooke described and illustrated the discoveries he had made using a compound microscope that he'd built. Hooke used the microscope to observe familiar objects such as insects, sponges, and feathers. When he put a thin slice of cork under the lens of his microscope, Hooke made a very important discovery. He saw the cell walls in the cork tissue. Hooke had discovered plant cells.

Even though his discoveries were amazing in his day, Hooke's microscope was quite crude. It didn't look that different from today's microscopes, but it had poorly ground lenses, which caused Hooke's view of the objects to be blurred or distorted. What's more, early microscopes could not magnify objects more than 20 or 30 times their actual size. By contrast, most microscopes found in secondary schools today can magnify objects up to 430 times.

LEEUWENHOEK PERFECTS THE LENS

Leeuwenhoek's major contribution to the development of the microscope was to make lenses that were much more finely ground than those used by Hooke and others. He never went to college, and he earned a living by selling fabric in a small shop. For him, making microscopes was a hobby that became a lifelong obsession.

Leeuwenhoek learned to grind lenses by observing the craftsmen who made eyeglasses in Delft. Leeuwenhoek's lenses, often no more than 0.3 centimeters (0.1 inches) across, were so even and perfect they provided clear images that were free of distortion. They could magnify objects to between 50 and 300 times their actual size. He mounted the tiny lenses in frames of gold and silver that he also crafted himself.

Unlike Hooke's compound microscope, Leeuwenhoek's device had only one lens. It was mounted in a tiny hole in a brass plate. Leeuwenhoek placed the object he wanted to examine on a sharp point in front of the lens. He adjusted the position with the screws. The entire device was less than 10 centimeters (3.9 inches) long.

READING SELECTION

EXTENDING YOUR KNOWLEDGE

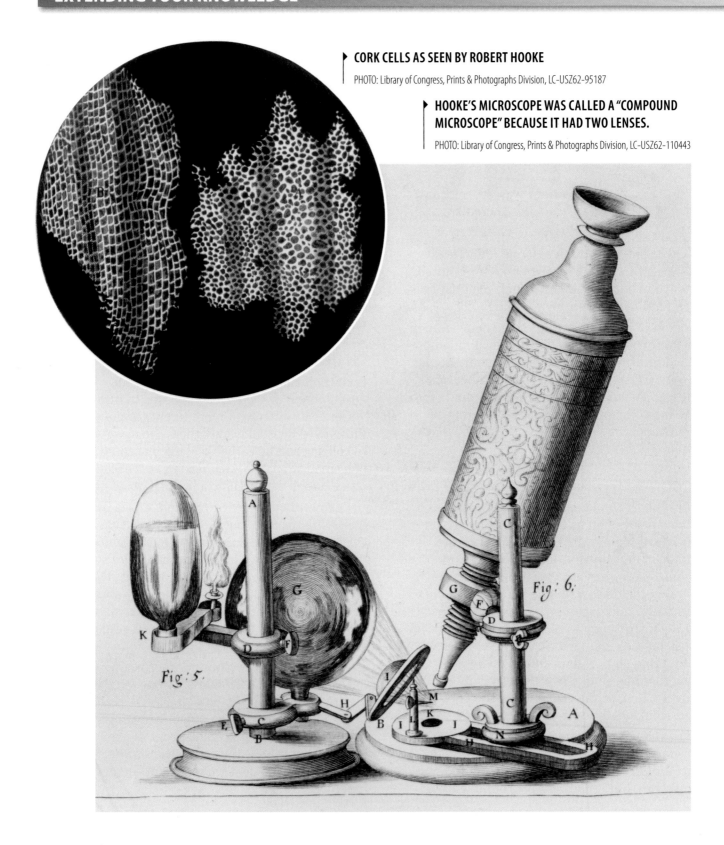

▶ **CORK CELLS AS SEEN BY ROBERT HOOKE**
PHOTO: Library of Congress, Prints & Photographs Division, LC-USZ62-95187

▶ **HOOKE'S MICROSCOPE WAS CALLED A "COMPOUND MICROSCOPE" BECAUSE IT HAD TWO LENSES.**
PHOTO: Library of Congress, Prints & Photographs Division, LC-USZ62-110443

For a scientist, good tools are just the start. Scientists also need the ability to observe carefully and to record their findings accurately. They need patience. Leeuwenhoek had all these qualities; in addition, he was very curious. He wrote about everything he saw, from algae on pond water to mineral crystals and fossils. He discovered microscopic organisms in rainwater. He discovered blood cells and was the first to see living sperm in an insect. He is credited with publishing the first drawing of bacteria.

Leeuwenhoek stuck just about everything under his lens—including plaque from his own teeth! What did he see? Something that wouldn't surprise your dentist at all. "I saw . . . many very little living animalcules," he wrote. "Very prettily a-moving. The biggest . . . had a very strong and swift motion . . . and shot through the water. The second . . . spun around like a top."

Hooke passed away in 1703, and Leeuwenhoek died in 1723, at the age of 91. Both had become world famous. Leeuwenhoek was so famous that Peter the Great, czar of Russia, once came to Delft to visit him at his home.

The science of microscopy has made great progress since the time of Hooke and Leeuwenhoek. To get an idea of how much progress, take a look at the image of a mite. It was taken through a scanning electron microscope that has a magnification range of 15 to 200,000 times! ■

▶ THIS MITE, WHICH MEASURES 150–200 MICRONS IN LENGTH (1/1000 MM), IS MAGNIFIED 850 TIMES ITS ACTUAL SIZE.

PHOTO: Photo by Eric Erbe, digital colorization by Chris Pooley, Agricultural Research Service/U.S. Department of Agriculture

DISCUSSION QUESTIONS

1. Early microscopists had to draw pictures of what they saw through the lenses if they wanted anyone else to see it. Modern microscopes are able to take digital images. Which kind of picture do you think gives a more accurate view of the object?

2. Why is it useful to be able to see tiny organisms so clearly?

INVESTIGATING *LUMBRICULUS*

INTRODUCTION

In this lesson, you will work with an organism called *Lumbriculus variegatus*, also known as the California blackworm. You will observe the structure of a blackworm with a hand lens and decide which familiar organism it most resembles. You will also try to identify its anterior (head) and posterior (tail) ends. You'll observe this creature more closely through the microscope and prepare a scientific drawing using the skills and techniques you learned in Lesson 2. You will observe blood pulsating through its blood vessels and measure its average pulse rate. Then you will be given a small fragment of a blackworm to observe and measure. You will place the fragment in a small plastic tube and observe it over three weeks for signs of change. Finally, you will update your group's organism photo card for the blackworm.

▶ **IF YOU LOOKED CAREFULLY THROUGH THE MUD IN THE SHALLOW AREAS OF THIS STREAM, YOU MIGHT BE SURPRISED AT THE NUMBER OF BLACKWORMS YOU WOULD FIND.**

PHOTO: National Science Resources Center

OBJECTIVES FOR THIS LESSON

▶ Observe, sketch, and measure a blackworm, and compare its structure to that of a related organism.

▶ Measure and record the average pulse rate of a blackworm.

▶ Make observations of a blackworm fragment each week for three weeks to look for signs of change.

▶ Update your organism photo card for blackworms.

▶ **MATERIALS FOR LESSON 3**

For you

Your copy of Student Sheet 2.3a: Guidelines for Scientific Drawings

1 copy of Student Sheet 3.1: Template for Blackworm Drawing

1 copy of Student Sheet 3.2: Average Pulse Rate of a Blackworm

1 copy of Student Sheet 3.3: Observations of Blackworm Fragments

For your group

1 set of organism photo cards
1 petri dish, lid or base
2 metric rulers, 30 cm (12 in.)
2 student timers (or clock with second hand)
2 blackworms
2 hand lenses
2 plastic pipettes
2 microcentrifuge tubes
1 pair of scissors
2 compound light microscopes
4 pieces of filter paper
2 plastic slides
1 black marker

GETTING STARTED

1 You will begin this lesson by working with your partner to make some brief observations of a blackworm. You will then work with your group and the class to answer several questions.

2 Work with your partner to prepare a petri dish to hold your blackworm. One pair of students will work with the top of a petri dish; the other pair will use the base. Each pair will use one piece of filter paper.

A. If your pair is using the petri dish lid, place the filter paper into the lid, but do not trim the paper.

B. If your pair is using the petri dish base, place it upside down on the filter paper. Trace a circle on the filter paper. Quickly cut out the circle and trim it so that it fits comfortably inside the dish. Place the circle in the dish.

3 Take your petri dish and filter paper to the blackworm culture container your teacher has prepared. Follow your teacher's directions to obtain a blackworm. Do not mix in water or chemicals from any other sources or you risk killing the blackworms.

4 Taking turns with your partner, use your hand lens to briefly observe the blackworm.

5 Work with your group to answer the following questions in your science notebook. Take another look at the blackworm if necessary. Share your answers with the class. ✐

A. What familiar organism does the blackworm resemble?

B. In what way(s) does the blackworm resemble this organism?

C. In what way(s) is it different from this organism?

D. Explain how you can tell the anterior end from the posterior end of your blackworm.

DRAWING AND MEASURING A BLACKWORM

PROCEDURE

1 If necessary, use your pipette to dampen the filter paper with a few drops of water from the culture container. Place the petri dish on the microscope stage and focus on the blackworm under a total magnification of 40x (see Figure 3.1).

2 Prepare a detailed drawing of your blackworm on Student Sheet 3.1: Template for Blackworm Drawing, as the student is doing in Figure 3.2. Follow the guidelines for scientific drawings from Student Sheet 2.3a. Title your drawing, "*Lumbriculus*: The California Blackworm."

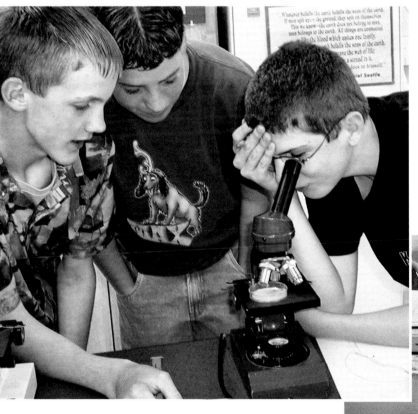

▶ THE TRANSLUCENT QUALITY OF THE BLACKWORM'S SKIN ALLOWS YOU TO SEE THE BLOOD VESSELS AND DIGESTIVE TUBE QUITE CLEARLY.
FIGURE **3.1**

PHOTO: Courtesy of Henry Milne/© NSRC

▶ REFER TO THE DIRECTIONS FOR SCIENTIFIC DRAWINGS WHEN YOU DRAW AND LABEL YOUR BLACKWORM.
FIGURE **3.2**

PHOTO: Courtesy of Henry Milne/© NSRC

Inquiry 3.1 continued

3 Label at least three structures of the blackworm on your drawing. Use the reading selection "More Than Just Bait" on pages 49-53 as a reference.

4 Follow these instructions to measure your blackworm:

A. Take out a piece of unused notebook paper and lay it on your desk.

B. Fill about one-third of your pipette with spring water from the blackworm culture container. Take it back to your desk and squirt the water into the petri dish containing the blackworm.

C. Tilt the petri dish to move the water around until the blackworm is soaked. Use your pipette to suck up the blackworm with a small amount of water.

D. Slowly and carefully squirt the water with the blackworm onto the notebook paper to form one large drop. Use the tip of the pipette to draw the water out in one direction. The blackworm will spread out to follow the trail of water, as shown in Figure 3.3.

E. Once the blackworm has spread out fully, measure it with the metric ruler. Be careful not to touch the blackworm or you may damage it. Record your measurement in the appropriate place on Student Sheet 3.1.

5 Use your pipette to suck up the blackworm from the notebook paper and squirt it back into the petri dish.

▶ **SLOWLY DRAW OUT THE DROP OF WATER USING THE TIP OF THE PLASTIC PIPETTE.**
FIGURE **3.3**

DETERMINING THE PULSE RATE OF A BLACKWORM

PROCEDURE

1 Use your scissors to cut a piece of filter paper into a rectangular shape that will fit on a microscope slide, as seen in Figure 3.4.

2 Use your pipette to obtain a few drops of water from the culture container. Place the filter paper on the slide, and use the water to dampen the paper. Using your pipette, transfer the blackworm from the petri dish to the slide. Place the slide on your microscope stage. Focus on a small section of the blackworm under the lowest power (40x). You should be able to see the blood moving through the vessels. Agree with your group on how to describe the movement of blood in your blackworm.

3 Pulse rate in blackworms can be defined as the number of pulsations of blood that pass by one location in 1 minute. Decide with your partner how you will measure the pulse rate of your blackworm. Think about the variables involved in this investigation. Discuss what you will keep constant each time you measure the pulse rate.

4 Set up a data table on Student Sheet 3.2: Average Pulse Rate of a Blackworm on which to record your data. Leave enough space for several trials. This will ensure more valid results.

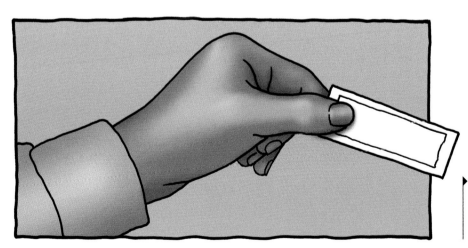

▶ MAKE SURE THE PIECE OF FILTER PAPER IS A LITTLE NARROWER THAN THE SLIDE.
FIGURE **3.4**

Inquiry 3.2 continued

5 Take turns with your partner counting the pulsations of blood and watching the timer or clock (See Figure 3.5). Record your information in your data table. After you have completed the counts, find the average pulse rate for your blackworm. If you have difficulty finding a pulse in your blackworm because it is moving around too much, ask your teacher to let you try a different blackworm that may be less active.

6 Follow your teacher's directions for returning your organism. Dispose of your filter paper, and return your petri dish and microscope slide as directed by your teacher.

▶ **THESE STUDENTS ARE WORKING TOGETHER TO MEASURE THE PULSE RATE OF THEIR BLACKWORM.**
FIGURE **3.5**

PHOTO: Courtesy of Henry Milne/© NSRC

INVESTIGATING REGENERATION OF BLACKWORMS

PROCEDURE

1 Use a marker to label a microcentrifuge tube with your group's name and class period. Use a plastic pipette to fill the plastic tube with water from the blackworm culture container. Snap the cap shut and set the tube aside.

2 Prepare a microscope slide and filter paper as you did in Procedure Steps 1 and 2 of Inquiry 3.2.

3 On Student Sheet 3.3: Observations of Blackworm Fragments, prepare a data table for recording your observations on four different dates over three weeks. Include space for recording the date, number of segments, and length of your blackworm fragment.

4 Take your slide to your teacher to obtain a blackworm fragment.

5 Use the lowest power and the dimmest light of your microscope to view your fragment, measure its length, and count the number of segments. Record your observations on the data table you prepared on Student Sheet 3.3. Be aware that too much heat, light, or excessive shaking could kill your blackworm fragment. Turn off the light immediately after counting the segments.

Inquiry 3.3 continued

6 When you have completed your measurements, use a pipette to transfer the fragment to the plastic tube. Snap the cover on the tube and give the tube to your teacher for storage.

7 Work with your group to update the blackworm organism photo card.

REFLECTING
ON WHAT
YOU'VE DONE

1 Based on what you have learned in this lesson, respond to the following in your science notebook:

A. Summarize the similarities and differences between blackworms and common earthworms using a Venn diagram.

B. What is a probable reason that one or both ends of some of your blackworms are lighter in color than the rest of their bodies?

C. Why do blackworms make some of their unusual movements?

D. What did you discover about the pulse rate of the blackworm when it was measured at different parts of its body?

E. If you measured your pulse at different places on your body, would you expect to observe the same pattern of results? Explain.

F. In addition to being fairly active, blackworms are much larger than organisms usually studied through the microscope. Because of this, blackworms have a more complex way of moving food and oxygen through their bodies. What evidence have you observed to support this statement?

G. What evidence did you observe that regeneration has occurred in your blackworm?

More Than Just Bait

What do you get when you cut a blackworm in half?

A. One dead worm

B. Two live worms

C. A bloody mess

Strangely, the answer is B. This amazing worm, whose scientific name is *Lumbriculus variegatus*, can be cut into several fragments—and it won't die or even bleed. Instead, it regenerates a new head or tail, or both, from the various pieces.

What's more amazing is that the blackworm is not a rare animal living in some faraway place. Usually no more than 10 centimeters (3.9 inches) long, this worm lives in the shallow edges of ponds, marshes, and lakes throughout North America and Europe.

Despite its short length, a mature blackworm has between 150 and 250 body segments. Even a fragment of blackworm only a few segments long can regenerate lost body parts—fast. In fact, fragmentation, followed by regeneration, is much more common than sexual reproduction in blackworms.

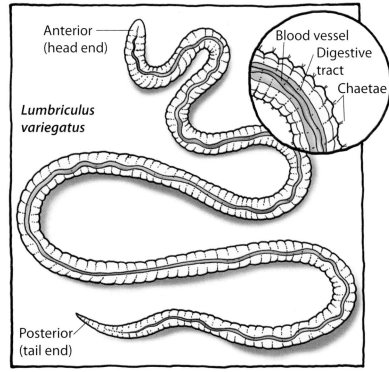

Lumbriculus variegatus

Anterior (head end)

Posterior (tail end)

Blood vessel
Digestive tract
Chaetae

▶ **BASIC ANATOMY OF AN ADULT BLACKWORM**

READING SELECTION
EXTENDING YOUR KNOWLEDGE

▶ **YOU CAN TELL THAT THE ANTERIOR END OF THIS BLACKWORM HAS UNDERGONE REGENERATION BECAUSE OF ITS PALE COLOR.**

PHOTO: William R. West/Carolina Biological Supply Company

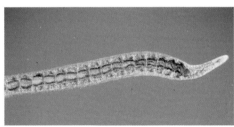

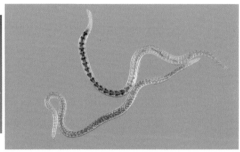

▶ **NOTE THE LIGHTER COLOR OF THE REGENERATED HEAD AND TAIL ENDS OF THESE BLACKWORMS.**

PHOTO: William R. West/Carolina Biological Supply Company

"The segments regenerate quickly," says Dr. Charles Drewes, a zoologist who has studied blackworms for many years. "For example, a new head or tail usually develops within two to three weeks. The new segments—usually eight for a head and between 20 and 100 for a tail—are smaller and paler than the original ones."

A WORM WITH A RAPID REFLEX

The blackworm "swims" by twisting its body through the water in a corkscrew fashion. If the water in which it lives is shallow enough, a blackworm will stretch its tail to the surface of the water. It then bends its tail at a right angle so that a few centimeters of its dorsal surface are lying just above the water's surface. Part of the tail now faces skyward and is exposed to air. Although this is a good position for gas exchange of oxygen and carbon dioxide, it exposes the blackworm's tail to its enemies.

To offset the problem of the tail's exposure, the worm uses a special rapid-escape reflex. The tail end rapidly shortens in response to a threatening enemy. This reflex can be triggered by touch, a vibration, or even by the sudden appearance of a shadow. Nerve cells called photoreceptors, which are able to detect these shadows, are located in the blackworm's tail.

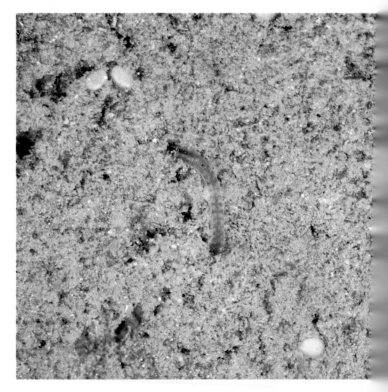

▶ **IF YOU LOOK CLOSELY, YOU CAN MAKE OUT THE TAIL OF A BLACKWORM BENT TO BE PARALLEL TO THE SURFACE OF THIS POND.**

PHOTO: Courtesy of Carolina Biological Supply Company

IT'S ALL IN THE FAMILY

If you haven't seen a blackworm in the wild, you've likely seen its relative, *Lumbricus terrestris*, the common earthworm. It, too, lives throughout North America and Europe—but in the soil. It can grow up to 25 centimeters (9.8 inches) long, and like the blackworm, it has the gift of regeneration.

A mature earthworm has about 150 segments. It also has a light-colored bulge on its body, called the clitellum. If an earthworm is cut in two, only the part with the clitellum can regenerate. The part without the clitellum will die.

▶ **THE BULGE AROUND THE EARTHWORM NEAR ITS CENTER IS CALLED THE CLITELLUM. IT PRODUCES MUCUS THAT FORMS A COCOON FOR THE WORM'S EGGS.**

PHOTO: Courtesy of Carolina Biological Supply Company

READING SELECTION

EXTENDING YOUR KNOWLEDGE

The next time you see an earthworm, look for its clitellum. Look even more carefully and you'll also see tiny hairs on each segment of its body. These hairs, called setae, help earthworms move by giving them many tiny grips on the soil. In blackworms, similar hairs are referred to as chaetae.

EARTH MOVERS

Earthworms have remarkable regeneration powers, and they are also terrific diggers. These mini-bulldozers actually plow and fertilize soil!

Here's how: First, they eat bits of soil, decaying leaves, bacteria, and other microorganisms. (Each bite enlarges their network of underground tunnels.) With digestive systems the length of their bodies, they next grind and mix their food. Then they expel their waste, called castings, which is actually first-rate, nutrient-rich soil. Throughout this process, these tiny farmers till the earth by bringing subsoil to the surface. That's not all! Their tunnels give air and water easy access to the roots of plants, helping them grow.

Both blackworms and earthworms are amazing animals that deserve our respect. So while we may have to look down to find them, we should never look down on them. ■

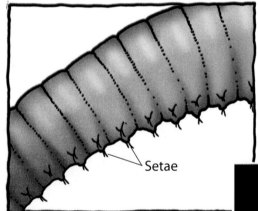

Setae

▶ **THESE TINY HAIRS HELP THE EARTHWORM CLING TO THE SOIL AS IT MOVES.**

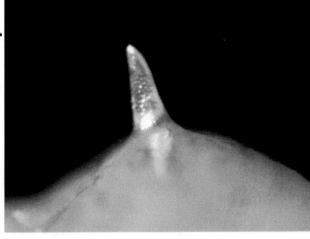

▶ **THIS MAGNIFIED PHOTO OF AN EARTHWORM'S SETA ALLOWS YOU TO SEE ITS ACTUAL STRUCTURE.**

PHOTO: Courtesy of Carolina Biological Supply Company

Are blackworms and earthworms the only two organisms that can undergo regeneration? Hardly! Another amazing regeneration story belongs to the starfish. A starfish can grow a new arm, or ray, if it loses one. A few starfish can even regenerate an entire body from a single ray. In some cases, several starfish can result from one starfish that gets cut into pieces.

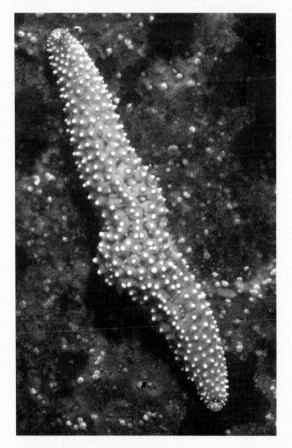

▶ THE SHAPE OF THIS STARFISH WILL BECOME MORE TYPICAL AS ITS PARTS REGENERATE FULLY.

PHOTO: Ed Bierman/creativecommons.org

 DISCUSSION QUESTIONS

1. How do the bodies and behaviors of blackworms and earthworms work together to help them survive in their habitats?

2. Suppose you wanted to know the size of the smallest blackworm fragment that could regenerate into a whole new blackworm. How could you find out?

LESSON 4

CREATING YOUR OWN POND

INTRODUCTION

An ecosystem is a community that includes living things and their environment, functioning together as a unit. There are many kinds of ecosystems. In this lesson, you will create your own pond ecosystem and begin observing the living and nonliving things found there. You will complete your observations in Lesson 8.

> PONDS CAN BE RECREATION AREAS, BUT THEY ARE ALSO ECOSYSTEMS.
>
> PHOTO: Robbie Sproule/ creativecommons.org

OBJECTIVES FOR THIS LESSON

▸ Construct a pond and observe, sketch, and label its layers.

▸ Observe and document the living things in the pond directly and with magnification.

▸ Explain the types of changes that may occur in your pond over a three-week period.

▸ MATERIALS FOR LESSON 4

For you

Your copy of Student Sheet 2.3a: Guidelines for Scientific Drawings

1 copy of Student Sheet 4.2: Sketches of Pond—Macro and Micro

For your group

1 clear plastic cup with lid
2 compound light microscopes
2 depression slides
2 coverslips
2 hand lenses
1 plastic pipette
2 pairs of scissors
1 small cotton ball
1 metric ruler, 30 cm (12 in.)
2 decaying leaves
2 pieces of hay
5 *Lemna* plants
5 grains of rice
1 black marker
1 box of colored pencils
Gravel
Soil
Spring water

GETTING STARTED

1 Work with your group to develop a list of at least six organisms you might expect to find in and around a pond. Write this list in your science notebook. 📝

2 Share your list with the class.

▶ AT CERTAIN TIMES OF THE YEAR, PONDS ARE
DOTTED WITH FROG EGG MASSES.

PHOTO: Pete Pattavina/U.S. Fish and Wildlife Service

CONSTRUCTING YOUR POND

PROCEDURE

1 Use the scoop to measure 50 cubic centimeters (cm³) of gravel into a beaker. (One cm³ is equivalent in volume to 1 milliliter [mL].) Pour the gravel into the plastic cup to form a layer on the bottom, as shown in Figure 4.1. Then, use the metric ruler and your marker to put a mark about 1.5 cm from the bottom of the cup.

2 Place soil on the gravel until it reaches that mark.

3 Use your scissors to cut the two leaves into smaller pieces and lay them flat on the surface of the soil and gravel.

4 Cut the hay into pieces about 5 cm long and place them on the leaves, as shown in Figure 4.2.

5 Gently pour into the cup approximately 350 mL of the water provided by your teacher.

▶ THE GRAVEL, AS WELL AS THE CUP BOTTOM, PROVIDES A BASE FOR YOUR POND.
FIGURE **4.1**

Lemna

Water

Hay

Leaves

Soil

Gravel

▶ LAY THE HAY ON TOP OF THE LEAVES. THEN ADD WATER.
FIGURE **4.2**

Inquiry 4.1 continued

INQUIRY 4.2

6 Use the tip of your pipette to transfer five *Lemna* plants from the culture container to your pond. Using your hand lens, count the number of leaves, called "fronds," on the five plants. Record the number of fronds in your science notebook. You will need the number for Lesson 8. ✑

7 Your final pond should look like the cup in Figure 4.2. Do not move the pond for several minutes. This will allow the soil to begin settling to the bottom.

8 Proceed to Inquiry 4.2 immediately after creating your pond.

OBSERVING YOUR POND

PROCEDURE

1 Observe your pond at eye level. In the box provided on Student Sheet 4.2: Sketches of Pond—Macro and Micro, sketch exactly what you see. Be very detailed about your observations. Label the layers you observe in the cup using the directions for scientific drawings from Student Sheet 2.3a. Color your drawing as accurately as you can.

2 Prepare and view a slide of water from your pond in the following manner:

A. Add several strands of cotton to the depression in your slide. This helps slow the movement of any microorganisms that are present.

B. Use a plastic pipette to obtain water from the bottom of your pond, just above the soil and gravel. Add one drop of pond water to the depression on the slide.

C. Place a coverslip over the drop of water by placing one edge of the coverslip onto the slide and lowering the other edge slowly to avoid trapping air bubbles beneath, as illustrated in Figure 4.3. This type of slide is called a wet mount.

D. Set the magnification to 100x; then move the slide around while you look for microorganisms through the eyepiece of your microscope. Sketch any microorganisms that you may see in the circles on your student sheet.

3 Repeat Procedure Step 2 with a water sample from the top level of your pond. You will make further observations when you revisit your pond in a later lesson, so be as thorough as you can for comparison purposes.

4 Use the marker to write your group members' names near the top of your cup. Add five grains of rice to your pond; then place the lid loosely on top. This will slow down the evaporation of water as well as expose the water to oxygen.

5 Follow your teacher's directions for storing the pond and cleaning up.

REFLECTING
ON WHAT
YOU'VE DONE

1 Look back at the list you generated in your science notebook during "Getting Started." Based on what you have observed, revise your list.

2 Following your list of pond organisms in your science notebook, predict the ways in which you think your pond will change over the next three weeks.

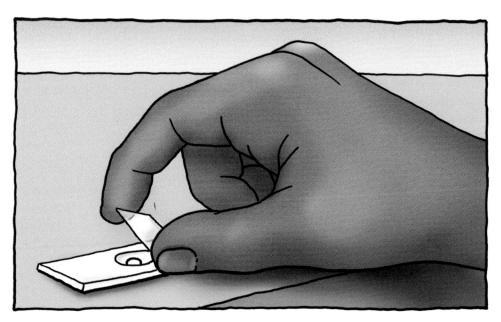

▶ PLACE ONE EDGE OF THE COVERSLIP DOWN FIRST.
FIGURE **4.3**

Even a Habitat Has to Live SOMEWHERE

What's the difference between a habitat and an ecosystem?

A habitat is the home of a single type of organism. A group of bats may have their habitat in a stand of trees, for instance. The branches meet their need for a roosting place to hang from, while the leaves may provide shelter from storms, predators, or hot sunlight.

Ecosystems are home to more than one type of organism. An ecosystem contains many habitats that may share resources such as water, land, and vegetation. The living and nonliving things in the ecosystem interact in various ways.

In the ecosystem containing a stand of trees, the bats eat the insects; the insects drink the sap from the trees; microorganisms in the soil decompose both the bats and the insects after they die. And

all these things—bats, insects, trees, water, soil, microorganisms—use air, light, and water. Trees use the light to photosynthesize; bats use the sunrise as a cue for sleep.

An ecosystem is defined by the interactions of organisms. In the ocean, which is a saltwater environment, bottle-nosed dolphins interact with squid—by eating them. Sea lions eat squid, too, so they compete with the dolphins. Since they all interact, bottle-nosed dolphins, squid, and sea lions are all part of the same ecosystem.

Freshwater rivers and streams, like the saltwater oceans, are aquatic (or water-based) ecosystems. Rainbow trout, which live in cool streams, share their ecosystem with plants and other organisms, including those they prey on, such as snails and young dragonflies. They don't share their ecosystem with adult dragonflies, who live in other, dry parts where they can warm their bodies.

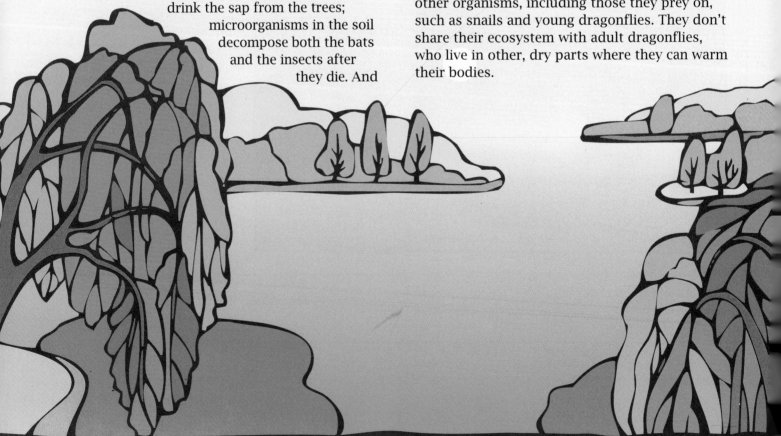

▶ **THIS POND IS HOME TO A GREAT DIVERSITY OF ORGANISMS.**

PHOTO: Courtesy of Henry Milne/© NSRC

An ecosystem can be as small as a puddle or as large as an ocean. An ecosystem can simply be a stretch of grassland or a rotten tree trunk. The size of an ecosystem is determined by the interactions between the organisms within it.

INSIDE A TYPICAL ECOSYSTEM—A POND

The living components of an ecosystem—plants, animals, fungi, and microorganisms—are referred to as a "community." A pond, for example, is an ecosystem in which a community of organisms, including blackworms, dragonflies, and *Lemna*, all interact among themselves and with their nonliving environment, such as the water, soil, and air.

The members of each species in a community are called a "population." All the blackworms in a single pond community, for instance, make up the blackworm population there.

READING SELECTION

EXTENDING YOUR KNOWLEDGE

Within a pond ecosystem are many different habitats, or homes. The population of blackworms may live in the muddy fringes of the pond. This is where they find shelter. This is also where they find food. This is their habitat—the place where their needs are met.

Other organisms that share the pond ecosystem may have different habitats. For example, *Lemna*, a plant also known as duckweed, lives on the water's surface, closer to the light from the sun. Crayfish, on the other hand, live at the bottom of the pond, scavenging for food that falls from upper layers.

Some organisms need to change their habitat as they grow up. Dragonflies and damselflies live the first part of their lives in the pond water habitat as nymphs. They eventually climb up plant stems, where they change into their adult form. Then the area above and around the pond becomes their habitat. It provides them with insects to eat, and with warmth, from sunlight.

FINDING AN ORGANISM'S NICHE

Organisms within an ecosystem perform certain jobs that keep the ecosystem functioning. In a pond, for example, birds and frogs keep the number of insects in check by eating them. In a grassy pasture habitat, dung beetles may eat the waste matter from cows and other animals, which helps to recycle nutrients. Although no organism applies for a job within an ecosystem, the things they do naturally become their jobs—their niches.

▶ THIS BALL OF DUNG, WHICH IS WASTE MATTER FROM ANOTHER ORGANISM, PROVIDES A FOOD SUPPLY FOR THIS BEETLE, APPROPRIATELY NAMED THE "DUNG BEETLE."

PHOTO: Andrew Ross/creativecommons.org

▶ THESE DAMSELFLY NYMPHS EVENTUALLY MOVE OUT OF THE WATER, UNLESS THEY BECOME FISH FOOD FIRST!

PHOTO: Courtesy of Carolina Biological Supply Company

THE ROLE OF NONLIVING THINGS

When we walk past a pond or through a stand of trees, we tend to look first for the living things—the plants and animals big enough for us to see. Often we don't notice the nonliving things in an ecosystem until they cause major changes. Air pollution might cause tree leaves to drop, for instance, and construction nearby might choke a stream with displaced soil, leaving the water stagnant and foul. During droughts, the soil gets parched and cracked, and we see all the living things react to the lack of water. Next time you go outside, notice the nonliving factors that contribute to the ecosystem around you, and think how that ecosystem might change if, say, your area received much more rain every year than it does now.

EVERYTHING CHANGES

Don't think for a minute that ecosystems, habitats, communities, and populations don't change. They do. Ponds dry up. Forests are ravaged by fires. Trees are blown down by hurricanes. Organisms become extinct. These are all natural processes. However, change also occurs because of human intervention. A river gets dammed, creating a lake in the process. Grasslands get mowed and turned into soybean fields. Or trees get cut down and replaced by parking lots or housing developments.

All over the world, animals, plants, and other species come and go—and habitats and ecosystems shift and change over time. Change, in fact, is one thing we can always rely on. ■

DISCUSSION QUESTIONS

1. Why are nonliving things included in the definition of "ecosystem"?

2. How many habitats can there be in a single ecosystem? Give examples.

▶ **THE DESTRUCTION OF THESE TREES GREATLY CHANGES THE NATURE OF THE ECOSYSTEMS AND HABITATS IN THIS AREA.**

PHOTO: NOAA's National Weather Service (NWS) Collection

EXPLORING CELLS

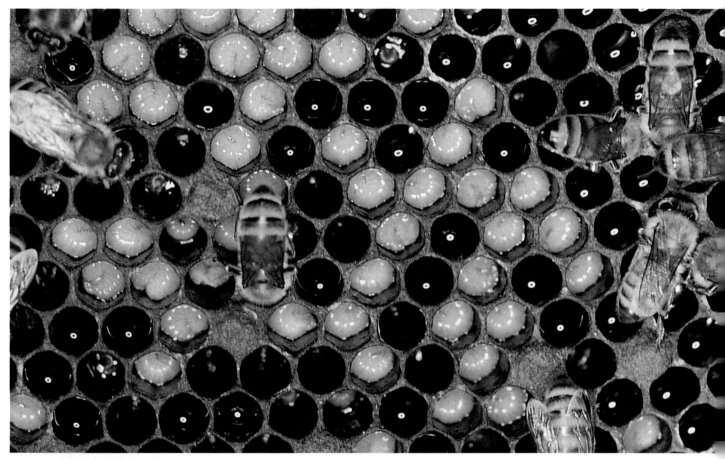

INTRODUCTION

In Lessons 1 through 4, you looked briefly at many of the organisms that you will explore in more depth in the rest of this unit. To learn more about organisms, you must first understand the nature of the cell, which is the basic unit of life. In this lesson, you will observe algal, plant, and animal cells through a microscope. You will draw, label, and measure the cells, following the guidelines for scientific drawings. You also will compare the structures of the cells and discuss whether their structures are suited to their functions.

▶ WHEN ROBERT HOOKE FIRST LOOKED AT A SLICE OF CORK THROUGH A MICROSCOPE, THE "TINY CAVITIES" HE DESCRIBED REMINDED HIM OF A BEE'S HONEYCOMB. IT PROMPTED HIM TO CALL THE TINY CAVITIES CELLS.

PHOTO: Courtesy of Carolina Biological Supply Company

OBJECTIVES FOR THIS LESSON

- Observe, draw, label, and measure cells based on specific guidelines.

- Observe and identify certain organelles of plant and animal cells.

- Observe the effect of salt solution on *Elodea* leaf cells.

- Compare the structure of various cells for evidence that they are suited to their functions.

- Update the organism photo cards for *Elodea*, *Spirogyra*, and humans.

MATERIALS FOR LESSON 5

For you

  Your copy of Student Sheet 2.3a: Guidelines for Scientific Drawings

  1 copy of Student Sheet 5.1: Template for *Spirogyra* Cell Drawing

  1 copy of Student Sheet 5.2: Template for Onion Leaf Cell Drawing

  1 copy of Student Sheet 5.3: Template for *Elodea* Leaf Cell Drawings

  1 copy of Student Sheet 5.4: Template for Animal Cell Drawings

  1 box of colored pencils

For your group

  1 set of organism photo cards

  1 sheet of newsprint

  2 pieces of sliced onion

  2 *Elodea* leaves soaked in fresh water

  2 *Elodea* leaves soaked in salt solution

  1 prepared slide of human cheek cells

  1 prepared slide of mammalian nerve cells

  2 compound light microscopes

  2 plastic slides

  2 coverslips

  2 metric rulers, 30 cm (12 in.)

  2 transparent rulers

  1 dropper bottle of Lugol solution

  1 plastic pipette

  Several strands of *Spirogyra*

GETTING STARTED

1. Work with your group to draw on a piece of newsprint what you think a typical cell looks like. Label any parts with which you are familiar.

2. Share your group's drawing with the class.

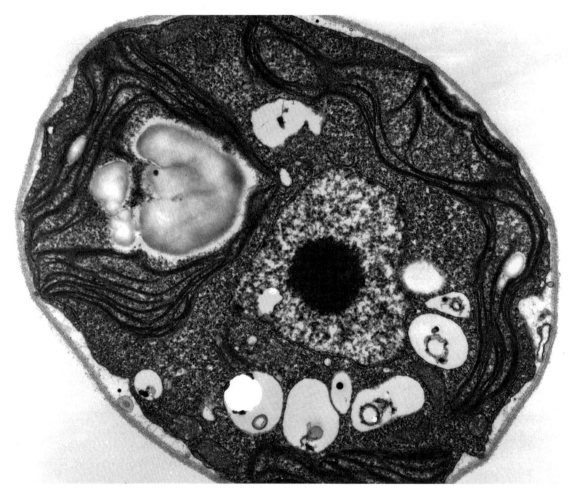

▶ **WHAT STRUCTURES DO YOU EXPECT TO FIND INSIDE A CELL?**

PHOTO: Dartmouth Electron Microscope Facility, Dartmouth College

OBSERVING, DRAWING, AND MEASURING AN ALGAL CELL

PROCEDURE

1 Read "Plant and Animal Cells: The Same, but Different" on pages 68-69. Discuss the reading selection with the class and ask questions to clarify anything you do not understand.

2 Use a plastic pipette to obtain a small sample of water from the container marked "*Spirogyra*." *Spirogyra* is a type of common pond alga whose cells are joined in chains. Make sure that the sample includes two to four of the green strands that are floating in the water.

3 Put a drop of the sample on the middle of the slide and add a coverslip.

4 Focus on a chain of *Spirogyra* cells under 100x. After observing *Spirogyra* through the microscope, discuss with your partner how you think it got its name.

5 Switch to 400x and focus on one cell (see Figure 5.1). Draw the cell in the circle on Student Sheet 5.1: Template for *Spirogyra* Cell Drawing. Title your drawing, "*Spirogyra* Cell." Label at least two organelles. Refer to "Plant and Animal Cells: The Same, but Different" to help you identify the structures. Follow the guidelines for scientific drawings on Student Sheet 2.3a.

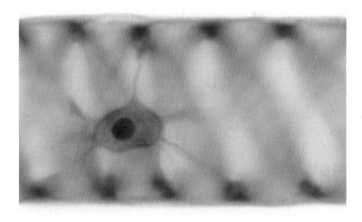

▶ **WITH A GOOD MICROSCOPE AND A LITTLE FINE TUNING, YOU CAN EVEN SEE THE NUCLEUS IN A *SPIROGYRA* CELL.**
FIGURE **5.1**

PHOTO: Courtesy of Carolina Biological Supply Company

READING SELECTION

BUILDING YOUR UNDERSTANDING

PLANT AND ANIMAL CELLS: THE SAME, BUT DIFFERENT

Almost all living things on Earth are made up of cells. Cells are the basic units of life. The simplest organisms—amoebae, for example—consist of only one cell. Complex organisms, such as humans, have trillions of cells that are divided into about 200 different types. Each cell type has a different function.

As the building blocks of living matter, plant and animal cells have many things in common. They also differ in some ways. Those differences are important because they point to some of the factors that distinguish one form of life from another. To understand this better, let's take a cross-sectional look at an animal cell and a plant cell.

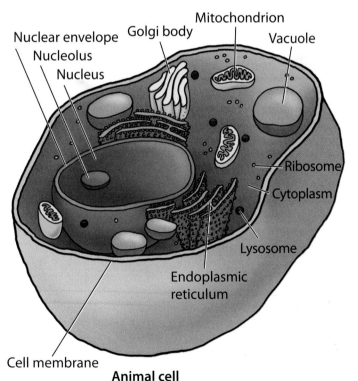

Nuclear envelope
Nucleolus
Nucleus
Golgi body
Mitochondrion
Vacuole
Ribosome
Cytoplasm
Lysosome
Endoplasmic reticulum
Cell membrane
Animal cell

INSIDE AN ANIMAL CELL

Although there is no typical animal cell, most animal cells have three basic parts that scientists call cellular organelles. "Organelle" means "little organ." The first organelle is the cell membrane, sometimes referred to as the plasma membrane. This living membrane separates the cell from the rest of its environment and helps control the passage of substances into and out of the cell.

The second cellular organelle is the nucleus, which usually occupies the central portion of the animal cell. Think of the nucleus as "command central." The nucleus regulates all the activities that take place in the cell. Instructions for the cell's activities are stored in the chromosomes, which are found in the nucleus. The chromosomes, which almost always occur in pairs, are composed of a substance called DNA (deoxyribonucleic acid). DNA carries the hereditary traits that are passed from parent to offspring. The nucleus is surrounded by a double membrane called the nuclear envelope.

The third basic cellular organelle is a jellylike substance called cytoplasm, which lies between the cell membrane and the nuclear envelope. In addition to the three basic cellular organelles, there are additional organelles in the cytoplasm. Each organelle carries out a specific cell function. For example, nutrients are broken down in the sausage-shaped organelles called mitochondria. The energy produced in these organelles is either released to support the cell's activities or stored in the cell for future use. That is why mitochondria are often referred to as the "powerhouses" of the cell.

Ribosomes are organelles that help make the proteins that the cell needs to perform its life activities. Many ribosomes are located along the endoplasmic reticulum, or ER for short. The ER is a series of cavities that is connected to the nuclear envelope. Some substances

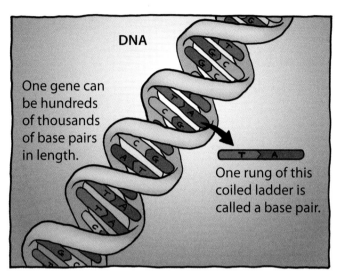

DNA

One gene can be hundreds of thousands of base pairs in length.

One rung of this coiled ladder is called a base pair.

▶ THIS IS A MODEL OF A PORTION OF A DNA MOLECULE THAT MAKES UP A CHROMOSOME. HEREDITARY STRUCTURES CALLED GENES ARE MADE UP OF VARYING NUMBERS OF BASE PAIRS.

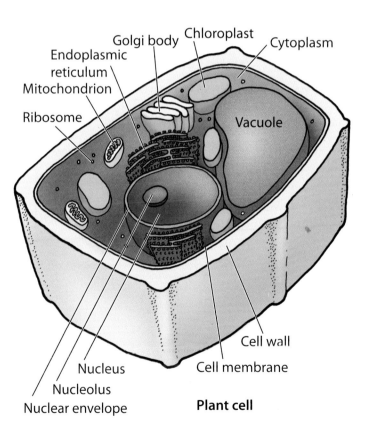

Endoplasmic reticulum
Golgi body
Chloroplast
Cytoplasm
Mitochondrion
Ribosome
Vacuole
Nucleus
Nucleolus
Nuclear envelope
Cell wall
Cell membrane

Plant cell

travel between the nucleus and cytoplasm through these cavities. Golgi bodies package the proteins made by the ribosomes so that they can be sent out of the cell. Organelles called lysosomes help the cell digest proteins. The cytoplasm also contains organelles called vacuoles. Filled with water, food, or waste, they are the cell's storage tanks.

INSIDE A PLANT CELL

Plant and animal cells have the same basic cell parts—cell membrane, nucleus, and cytoplasm. But there are some differences.

First, the plant cell is surrounded by a rigid, outer layer called the cell wall. The cell wall contains cellulose, a tough substance that supports and protects the cell. Like the cell membrane that lies within, the cell wall allows materials to pass into and out of the cell. Unlike the cell membrane, the cell wall is nonliving.

The nucleus is much the same in plant and animal cells. But some of the organelles in the plant cell's cytoplasm are different. For example, some plant cells have organelles called plastids, which contain pigments. Pigments give parts of plants their characteristic colors—red for tomatoes, orange for carrots, and green for spinach.

A chloroplast is a special plastid in a plant's leaf and stem cells. Chloroplasts contain a green pigment called chlorophyll. Chlorophyll traps energy from the sun. Plant cells use this energy to produce glucose, a simple sugar, during a process called photosynthesis. The vacuoles in plant cells are much larger than those in animal cells. Most plants have a large central vacuole that helps support the plant cell and also serves as a storage place for water, sugar, starch, and protein.

Although plant and animal cells have many organelles in common, each has organelles that the other does not, making them the same, but different! ■

Inquiry 5.1 continued

6. Using your transparent ruler as a coverslip, measure the length of one *Spirogyra* cell, as seen in Figure 5.2. Record the measurement in parentheses to the right of the title below the circle.

7. Review your drawing to ensure that you have followed each guideline. Follow your teacher's directions for turning in your drawing.

8. Follow your teacher's directions for putting away your supplies and cleaning up.

INQUIRY **5.2**

OBSERVING, DRAWING, AND MEASURING AN ONION LEAF CELL

PROCEDURE

1. Follow these steps to prepare a wet-mount slide of a leaf cell from the bulb of an onion plant:

   A. Obtain a piece of sliced onion from the container provided by your teacher.

   B. Follow the steps in Figure 5.3 to prepare a wet-mount slide of an onion leaf.

SAFETY TIP

Wear splash-proof safety goggles whenever you use chemicals such as Lugol solution. Lugol solution will stain skin and clothing and can be harmful when it comes in contact with your eyes or mouth.

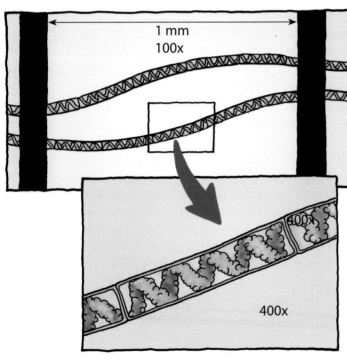

1 mm
100x

400x

▶ MEASURING THE LENGTH OF THE CELL USING THE TRANSPARENT RULER
FIGURE **5.2**

2 Take turns with your partner to prepare a drawing of one onion leaf cell under high magnification. Draw the cell in the circle on Student Sheet 5.2: Template for Onion Leaf Cell Drawing. Follow the guidelines for scientific drawings. Title your drawing "Onion Leaf Cell." Label the cell wall, nucleus, and cytoplasm. Discuss with your partner why you cannot see chloroplasts in these cells even though these are plant cells.

3 Use the transparent ruler to measure the length of the cell. Record the measurement in the appropriate place on your drawing.

4 When you and your partner have completed your drawings, rinse and dry your slide, coverslip, and transparent ruler, and proceed to the next inquiry.

1. Put one drop of Lugol solution in the middle of a plastic slide.

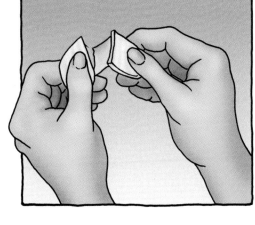

2. Bend the piece of onion against the curve until it snaps. Push one side under the other to peel off a thin membrane of leaf epidermis.

3. Spread the membrane out in the Lugol solution so that it is flat on the slide. Add a coverslip.

▶ HOW TO PREPARE ONION LEAF MEMBRANE FOR VIEWING UNDER THE MICROSCOPE
FIGURE **5.3**

INQUIRY 5.3

OBSERVING, DRAWING, AND MEASURING *ELODEA* LEAF CELLS

PROCEDURE

1 Follow your teacher's directions to obtain one *Elodea* leaf that is soaking in fresh water. *Elodea* is a common freshwater plant whose leaves are good specimens for observing typical plant leaf cells.

2 Place the leaf on the slide and add a coverslip. Focus on a layer of cells under 100x. Switch to 400x and focus on a smaller group of cells. With your partner, discuss which structures you can see in these cells that were not present in the onion leaf cells.

3 Draw one cell in the left-hand circle on Student Sheet 5.3: Template for *Elodea* Leaf Cell Drawings. Title your drawing "*Elodea* Leaf Cell." Label three of its organelles. Use the transparent ruler to measure the length of one cell. Record the length in the appropriate place on your drawing.

4 Clean your slide and then obtain a second *Elodea* leaf, which has been soaking in salt water. Set up your slide as you did for the *Elodea* that had been soaking in fresh water. Move the slide around to find a cell whose contents have shrunk into a round or oval shape. Draw one cell under 400x in the right-hand circle on Student Sheet 5.3. Label the same three organelles as you did in Procedure Step 3. Also label a fourth organelle that has become visible because water has been drawn out of the cell by the salt solution.

5 Rinse and dry your slide, coverslip, and transparent ruler.

EXPLORING
ANIMAL CELLS

PROCEDURE

1 Have one member of your group obtain one prepared slide of mammalian epithelial tissue (cheek cells) and one of mammalian nerve tissue.

2 Take turns with your partner to observe, draw, and measure one cell from one of the prepared slides. When each pair in your group is finished with its drawing, trade slides. Draw one slide in each of the circles on Student Sheet 5.4: Template for Animal Cell Drawings.

3 Title the appropriate drawing "Cheek Cell." Label the cell membrane, cytoplasm, and nucleus.

4 Title the other drawing "Nerve Cell." Label the cell membrane, cytoplasm, and nucleus. Refer to Figure 5.4 if you have difficulty identifying a cell within the nerve tissue.

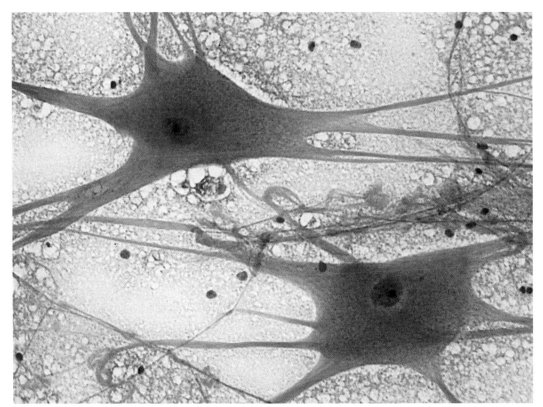

▶ ANIMAL CELLS ARE OFTEN DIFFICULT TO DISTINGUISH ON A SLIDE. THIS PHOTO SHOULD HELP YOU WITH YOUR IDENTIFICATION. THE CELL MEMBRANE, CYTOPLASM, AND NUCLEUS IN THESE TWO NERVE CELLS ARE CLEARLY VISIBLE.
FIGURE **5.4**

PHOTO: Courtesy of Carolina Biological Supply Company

Inquiry 5.4 *continued*

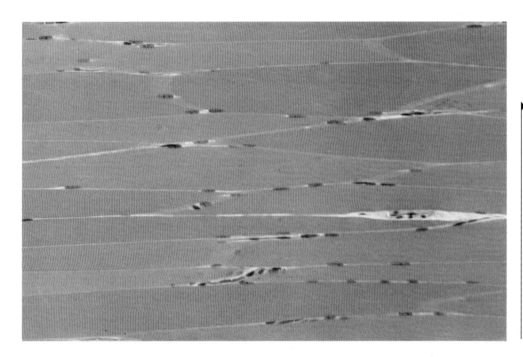

▶ THE INDIVIDUAL CELLS IN THIS PHOTO OF SKELETAL MUSCLE— TAKEN THROUGH A MICROSCOPE AT APPROXIMATELY 400X—ARE LONG, NARROW, AND SO TIGHTLY PACKED THAT THEY ARE DIFFICULT TO IDENTIFY.
FIGURE **5.5**

PHOTO: Courtesy of Carolina Biological Supply Company

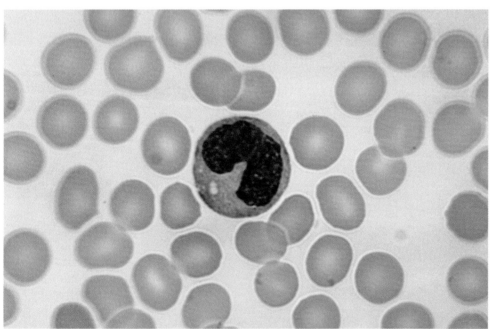

▶ THIS PHOTO OF HIGHLY MAGNIFIED BLOOD TISSUE CONTAINS MANY RED BLOOD CELLS AND ONE WHITE BLOOD CELL, RIGHT IN THE CENTER.
FIGURE **5.6**

PHOTO: Courtesy of Carolina Biological Supply Company

5 Look at the cells in your drawings and those in Figures 5.5 and 5.6. Discuss with your partner why these cells are so different in size and shape.

6 Work with others in your group to update your organism photo cards for *Spirogyra*, *Elodea*, and humans. Return them to your teacher.

REFLECTING
ON WHAT
YOU'VE DONE

1 Answer the following questions in your science notebook:

A. Based on your observations of algal *Spirogyra*, would you consider *Spirogyra* to be more plant-like or animal-like? Defend your answer.

B. Why do you think the bulb of the onion plant is so big? What function does it serve?

C. What happened to *Elodea* leaf cells when they were soaked in salt solution? How do you think this relates to what happens when you eat salty foods?

D. Use a Venn diagram to show the cell structures and organelles you observed in the onion bulb, *Elodea* leaf, and epithelial tissue.

E. Animal cells don't have cell walls. What gives animals such as mammals shape and support?

F. Give one example of how the size and shape of a cell is well suited for its particular function based on what you observed in this lesson.

G. Compare the cells you drew in the inquiries with your group's sketch from "Getting Started." Based on what you've learned, discuss how to make the sketch more accurate.

Who Eats Whom?

HETEROTROPHS MEET AUTOTROPHS

An ecosystem contains many distinct populations of organisms, and they all have to eat. Some of them even eat each other. But who eats whom? Populations of organisms can be categorized by how they get their food. Some organisms can make their own food, while others cannot.

▶ WHO IS MAKING FOOD IN THIS PICTURE?

PHOTO: chadh/creativecommons.org

ORGANISMS THAT MAKE THEIR OWN FOOD

Autotrophs are organisms that can make their own food. Where did they get their name? "Auto" means "self" and "troph" means "food." Plants, algae, and some bacteria make food for themselves using sunlight, water, and carbon dioxide; they are self-feeding. Autotrophs don't have mouths, because they don't need to eat anything at all. Wouldn't it be handy if you never had to stop playing a game to eat dinner?

Autotrophs do have to work for their food, though. Making food with the energy of sunlight, or photosynthesis, means using small but quite powerful factories inside their cells to dismantle carbon dioxide and turn it into sugars, fats, and all the other molecules they need. These molecules, especially the sugars, are the "food" that fuels all life on Earth. What this means is that all "food" is really stored energy from the sun.

When we look at autotrophic organisms in an ecosystem, we call them "producers." That's because they produce the food for all organisms that cannot make their own food. Every time a giraffe chews on some leaves, it has the tree to thank for providing its meal. Every time a butterfly sips some nectar, it has the plant to thank for providing its meal.

ORGANISMS THAT HAVE TO EAT FOOD

Heterotrophs get their food by feeding on other organisms. They cannot make food themselves. Where did they get their name? Combine "hetero," which means "different," with "troph," which, you'll remember, means "food." Heterotrophs must eat other organisms.

In an ecosystem, heterotrophs are called "consumers" because they consume food, rather than produce it. All animals, including humans, are heterotrophs. So are fungi and most bacteria. Although they are so small that you cannot see bacteria feeding, you can see the evidence of their eating. Yogurt, for instance, is the product of bacteria that eat milk. These bacteria, which live in the milk, eat the milk protein and excrete a material that makes the milk thicken.

All heterotrophs need to eat, but they do not all eat alike. Some are vegetarians who only eat plants (herbivores). Others like to eat both plants and animals (omnivores). Still others eat only animals (carnivores).

Specializing in particular foods is one of the things that defines an organism's way of life. Organisms that are carnivores, for example, have to find, catch, and kill their food. They develop adaptations to be successful at this challenging task. Some pit-viper rattlesnakes have pits on either side of their head that are able to sense heat. When a mouse or other warm-blooded animal scurries by, they detect and catch it. The venom that they inject with their fangs makes the animal become sick and weak so that they can swallow it whole. A pit viper would be great for pest control in your home, but would you want to share your home with one?

▶ **DESPITE THEIR DIFFERENCE IN SIZE AND APPEARANCE, ALL OF THESE CREATURES ARE HETEROTROPHIC.**

PHOTO (right): James Gathany, CDC/Dr. Christopher Paddock
PHOTO (top left): foshie/creativecommons.org
PHOTO (bottom): Frank Kovalchek/creativecommons.org

READING SELECTION
EXTENDING YOUR KNOWLEDGE

▶ **THIS MILLIPEDE IS AN EXPERT AT BREAKING DOWN LEAF LITTER ON THE FOREST FLOOR BY EATING IT.**

PHOTO: Axel Rouvin/creativecommons.org

An important type of heterotroph is the decomposer, which feeds on dead organisms and waste materials. Millipedes, mushrooms, molds, and bacteria are all examples of decomposers. By feeding on waste and the dead, decomposers help to break them down and return their molecules to the ecosystem, enriching the soil in which autotrophs grow. You could compare decomposers to garbage disposals. Without decomposers, dead organisms and waste would pile up.

RELATIONSHIPS

In any ecosystem, autotrophs and heterotrophs have a complex web of relationships based on who eats whom—we call these "feeding relationships." Most autotrophs are eaten by many different kinds of heterotrophs, who are eaten in turn by other heterotrophs.

Consider a grassland ecosystem. Feeding relationships exist even in the soil. The grasses and other plants are the autotrophs that capture energy from the sun and make food in their roots and shoots. Heterotrophs such as bacteria, fungi, and little root-eating worms called nematodes feed on the roots and shoots. Other small heterotrophs, such as insects, feed on these heterotrophs, and finally, larger animals such as birds and mammals take their turn. What a feeding frenzy is started by those plants!

Could heterotrophs survive without autotrophs? It would be tough. Try to imagine

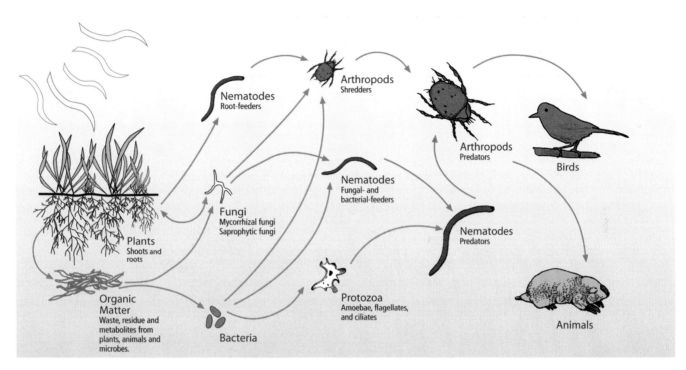

▸ **SOIL FOOD WEB**

PHOTO: USDA Soil Biology Primer

what you would eat if there were no autotrophs in the world. Is there any food that you eat that does not rely on autotrophic organisms? Even the meat in a hamburger comes from a cow that relies on grasses.

What about autotrophs existing without heterotrophs? Yes, indeed. It is believed that the first organisms on Earth were tiny autotrophs. Their ability to photosynthesize allowed the rest of life on Earth to evolve. Several billion years ago, this remarkable achievement of an organism being able to make its own food changed the course of history on our planet. Without autotrophs, we'd be history ourselves! ■

DISCUSSION QUESTIONS

1. Could a grassland ecosystem function if either the autotrophs or heterotrophs were removed? Explain.

2. Describe how autotrophs and heterotrophs complement each other in terms of inputs and outputs.

EXPLORING MICROORGANISMS

INTRODUCTION

In Lesson 4, you created your own pond ecosystem. Soon you will revisit your pond to observe any new developments. This lesson will prepare you to make those observations. During this lesson, you will observe four types of microorganisms and decide whether their characteristics are more animal-like or plant-like. You also will draw and label the microorganisms and estimate their lengths. You will create a cartoon featuring one of the microorganisms you observe. You will learn about the effects that microorganisms have had on our world. Finally, you will read about organisms whose benefits to humans are often misunderstood.

▶ **USING A MICROSCOPE, YOU'LL OFTEN SEE A VARIETY OF ORGANISMS IN JUST ONE DROP OF WATER.**

PHOTO: Charlotte Raymond, Photographer

OBJECTIVES FOR THIS LESSON

- Make a list of things you already know about microorganisms.

- Observe four species of living microorganisms called protists and identify their animal-like and plant-like characteristics.

- Observe, draw, and estimate the length of four protists.

- Create a cartoon using an *Amoeba*, *Euglena*, or *Paramecium* as the main character.

- Update your organism photo cards for *Amoeba*, *Euglena*, and *Paramecium*.

▶ **MATERIALS FOR LESSON 6**

For you

Your	copy of Student Sheet 2.3a: Guidelines for Scientific Drawings
1	copy of Student Sheet 6.2: Template for Protist Drawings
1	box of colored pencils

For your group

2	copies of Inquiry Master 6.3b: Scoring Rubric for Protist Cartoons
1	set of organism photo cards
2	compound light microscopes
2	depression slides
2	coverslips
2	transparent rulers
2	metric rulers, 30 cm (12 in.)
8	strands of cotton
1	*Amoeba*
1	*Paramecium*
1	*Euglena*
1	*Volvox*
1	*Amoeba* slide
1	*Paramecium* slide
1	*Euglena* slide
1	*Volvox* slide

GETTING STARTED

1. With your group, list in your science notebook five things you already know about microorganisms. ✍

2. Discuss your list with the class.

3. With your class, read the Introduction and "The Fine Art of Naming Organisms."

READING SELECTION

BUILDING YOUR UNDERSTANDING

THE FINE ART OF NAMING ORGANISMS

One of the first microorganisms that scientists viewed through a microscope was a squirmy little creature they named "amoeba." Latin and Greek were the languages used by scientists in the Western world at that time, late in the 17th century. The name "amoeba" is based on the Latin and Greek words for "to change." Scientists thought the name was appropriate because the amoeba's shape was always changing.

Although the meaning of "amoeba" is pretty straightforward, learning how to write the word properly can be confusing. For example, because the English language does not contain the Greek sound represented by the letters "oe," the English spelling of amoeba has never been consistent. Depending on what you are reading, you may see this word appear as *Amoeba*, *Amaeba*, *Ameba*, *amoeba*, or *ameba*!

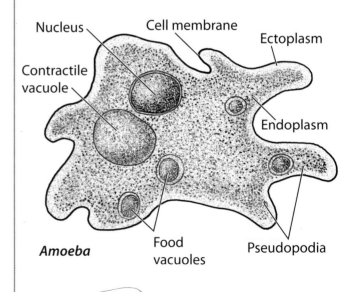

Amoeba

Adding to the confusion is the fact that the genus name *Amoeba*, which is capitalized and italicized, has also become this organism's common name, amoeba, which is neither capitalized nor italicized. Furthermore, when talking about more than one of these microorganisms, some people use the Latin plural form, amoebae, but others use the English plural, amoebas.

Another microorganism named by early scientists was Paramecium. Because it is shaped like a slipper, they named it using the Greek word for "oval." People now use "paramecium" (no capital letter, no italics) as the common name for this organism. The most common plural form is "paramecia." Paramecia are found in fresh water around the world. They are among the most complex of all single-celled organisms.

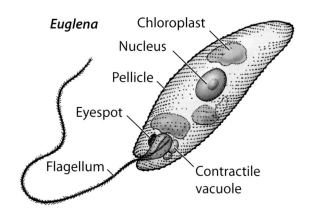

Euglena

- Chloroplast
- Nucleus
- Pellicle
- Eyespot
- Flagellum
- Contractile vacuole

scientists gave it the scientific name *Euglena*, from the Greek words that mean "true pupil of the eye." *Euglena* is found in ponds and pools of water. It is especially common in waters rich in chemicals.

The twisting, turning spheres that scientists named *Volvox* (from the Latin verb "to roll") seemed part animal, part plant. For many years, scientists classified *Volvox* as an animal. *Volvox* and the other microorganisms described above are now referred to informally as protists, a term used to describe eukaryotes (organisms with cells that contain a nucleus and other organelles enclosed by membranes) that are not plants, animals, nor fungi. Protists used to have their own kingdom, but scientists have since discovered that some protists are more closely related to plants, animals, and fungi than they are to each other. As a result, protists are now classified in several different kingdoms. ■

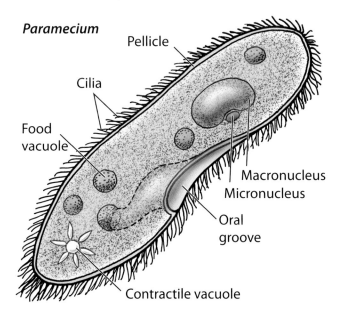

Paramecium

- Pellicle
- Cilia
- Food vacuole
- Macronucleus
- Micronucleus
- Oral groove
- Contractile vacuole

A third freshwater organism named by early scientists seemed like a cross between an animal and a plant. Like many plants, it was bright green. But it moved like an animal, and it did not have a cell wall. Impressed by this microbe's ability to use its tiny eyespot to find the brightest areas in its environment,

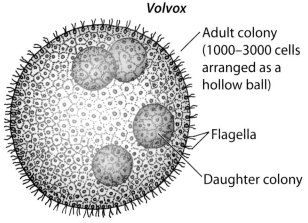

Volvox

- Adult colony (1000–3000 cells arranged as a hollow ball)
- Flagella
- Daughter colony

INQUIRY  6.1

EXPLORING LIVING PROTISTS

PROCEDURE

1 Review the following steps to obtain, observe, and record information about the four living protists your teacher has made available. You may observe the protists in any order you wish.

2 On a new page in your science notebook, create a table like Table 6.1 to record observations about the protists. Make your table the size of a full page. ✐

3 Have one student from your pair take a slide and coverslip to the center where your teacher has placed the containers of protists, the plastic pipettes, and cotton balls. Place two or three strands of cotton in the depression on the slide. The cotton helps confine the protists to a smaller area.

4 Use a plastic pipette to obtain a sample of water from one of the culture containers. Check to see if there are any special instructions for obtaining the protists. The instructions will be on the label of the container or on a nearby card.

5 Transfer the protists to the slide by squeezing a single drop of water from the pipette into the slide's depression. Add a coverslip. Return to your seat and place the slide on the microscope stage.

6 With your partner, locate the protist under a magnification of 100x. If you cannot find the protist, clean your slide, then go back and obtain another drop of water. Center one of the protists in the field of view and switch to a magnification of 400x, as seen in Figure 6.1. If you cannot keep the microbe in the field of view under 400x, switch back to 100x.

TABLE 6.1 OBSERVATIONS OF MICROORGANISMS

PROTIST	ANIMAL-LIKE FEATURES	PLANT-LIKE FEATURES	STRUCTURES/METHODS OF MOVEMENT
AMOEBA			
PARAMECIUM			
EUGLENA			
VOLVOX			

7 With your partner, decide which of the protist's features and behaviors are animal-like and which are plant-like. List these features in the table.

8 Try to determine how the protist moves. Identify any structures that seem to help it move.

9 Identify as many of the organelles as you can that are labeled in the illustrations in "The Fine Art of Naming Organisms."

10 When you have finished observing a protist, fill a plastic pipette with water from the appropriate culture container. Hold the slide over the water in the culture container while you use the plastic pipette to squirt the water from the container over the slide, as shown in Figure 6.2. This should wash the protists back into the culture container. Be careful not to allow cotton fibers to get into the container.

11 Rinse the slide and coverslip with tap water. Lay them on a dry paper towel and flip them over several times until they are dry. Continue working with the other protists until you are finished.

▶ THIS STUDENT IS USING THE FINE ADJUSTMENT KNOBS TO FOCUS ON A MICROBE UNDER HIGH MAGNIFICATION.
FIGURE **6.1**

PHOTO: Courtesy of Henry Milne/© NSRC

▶ HOLD THE SLIDE OVER THE CULTURE CONTAINER WITH ONE HAND WHILE SQUIRTING WATER OVER THE TOP OF THE SLIDE WITH A PIPETTE IN THE OTHER HAND.
FIGURE **6.2**

INQUIRY **6.2**

OBSERVING AND DRAWING PROTISTS FROM PREPARED SLIDES

PROCEDURE

1 Review the steps below to prepare scientific drawings from prepared slides of the four protists—*Amoeba*, *Paramecium*, *Euglena*, and *Volvox*.

2 Take turns with your partner to observe each of the four protists and draw each on Student Sheet 6.2: Template for Protist Drawings. Use the magnification that allows you to see the structures of the protist most clearly. You may draw the protists in any order. Follow the guidelines for scientific drawings on Student Sheet 2.3a. Title each drawing with the appropriate name. Label all of the organelles you can identify. Use your metric ruler to draw the lines for labels.

3 As you did in Lesson 2, use the transparent ruler to measure the diameter of the field of view at both 100x and 400x. Use those measurements to help you estimate the length of each of the four protists. Record the estimated lengths in millimeters in the appropriate place on Student Sheet 6.2.

4 As you complete each drawing, trade slides or switch seats with a pair of students who have finished using a slide of a different protist.

5 When you have completed your drawings, follow your teacher's directions for cleaning up.

6 With your group, update your organism photo cards for *Amoeba*, *Euglena*, and *Paramecium*. (Note that there is no organism photo card for *Volvox*.)

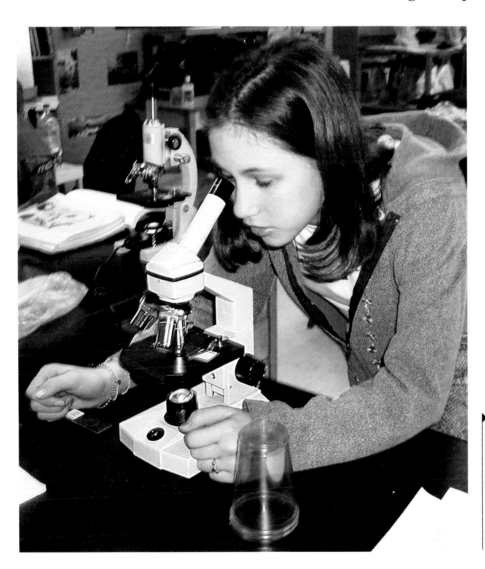

▶ STUDENTS HAVE DIFFERENT STYLES WHEN WORKING WITH THE MICROSCOPE. NOTICE THAT THE MICROSCOPE IN THIS PHOTO IS IN A DIFFERENT POSITION THAN THE ONE IN FIGURE 6.1.
FIGURE **6.3**

PHOTO: Courtesy of Henry Milne/© NSRC

CREATING A PROTIST CARTOON

PROCEDURE

1 Develop an idea for a cartoon with either an *Amoeba*, a *Euglena*, or a *Paramecium* as the central character. Your class will review examples of cartoons created by other students.

2 Your cartoon must be accurately drawn and should show details of at least four organelles, such as flagella, cilia, pseudopods, nuclei, vacuoles, cell membranes (or pellicles), contractile vacuoles, oral grooves, and eyespots (stigmas).

3 Make your caption humorous, focusing on at least one major characteristic of the protist.

4 Make a draft copy in pencil; use colored pencils for your final version. You will complete this assignment at home. If available, you may use a computer to create a background and labels for your final cartoon. Your teacher will tell you when your cartoon is due.

5 Use the rubric on Inquiry Master 6.3b: Scoring Rubric for Protist Cartoons to evaluate your cartoon before you turn it in. Your teacher will use this rubric, or one similar to it, to assess your work and will inform you of the point values for each box within the rubric.

REFLECTING
ON WHAT
YOU'VE DONE

1 Based on what you have learned in this lesson, respond to the following in your science notebook:

A. What are some of the characteristics you noticed while observing the four protists?

B. Why do you think protists are not classified as animals or plants?

2 Revisit the list you developed during "Getting Started." Discuss with your group what should be revised.

3 Based on what you learned in the reading selections "Mighty Microbes" on pages 93–95 and "Bacteria: Friends or Foes" on pages 90–92, answer the following questions in your science notebook:

A. Why is the misuse of antibacterial products and antibiotics potentially dangerous?

B. What are at least four ways in which microorganisms are beneficial to humans?

BACTERIA:
FRIENDS OR FOES?

You can't get away from bacteria. They're everywhere—in the soil, air, water, and even inside rocks. They're also in plants and animals. If you put all the world's bacteria in one pan of a scale, and the rest of life in the other, the bacteria's side would be heavier. Scientists estimate that there are 60,000,000,000,000,000,000,000,000,000,000 bacteria alive in the world today. This means that there are far more bacteria than people, which number around 7,000,000,000.

Bacteria are classified in their own domain, called Bacteria, apart from animals, plants, and fungi. They are an extremely diverse group of organisms. In your mouth alone live more than 400 different species of bacteria. Organisms that were once classified as blue-green algae, such as *Oscillatoria* and *Spirulina*, are now considered to be bacteria because scientists have found that genetically, they are more similar to bacteria than to algae.

Without a microscope, you can't see individual bacteria cells. Yet despite their small size, bacteria aren't all alike. They come in three main shapes—rod, spiral, and spherical.

We often name them by their shapes. Spherical bacteria are called *coccus*; rod-shaped bacteria are called *bacillus*; and spiral bacteria are called *spirilla*.

We all know that certain bacteria can make us sick. For example, strep throat is caused by a bacterium called *Streptococcus pyogenes*. (Can you guess the shape of this bacterium?) Because of this and their very small size, bacteria aren't normally used as specimens in secondary-school classrooms. To see bacteria in detail requires much more powerful microscopes, such as electron microscopes.

Did you also know that there are many more harmless—and even beneficial—bacteria than there are harmful ones? That's right! Some kinds of bacteria live in our digestive system. They help us process our food. Some of the intestinal gas we find so offensive forms when bacteria feed on undigested waste in our large intestine.

Another kind of bacteria lives in the roots of peas, beans, and other plants. These bacteria help put nitrogen into the soil. Without nitrogen, many plants would not survive. In fact, soil is loaded with bacteria, which keep nutrients

Rod

Spherical

Spiral

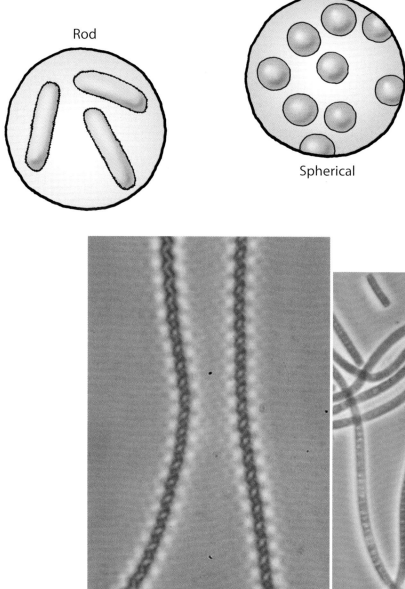

OSCILLATORIA

PHOTO: Courtesy of Carolina
Biological Supply Company

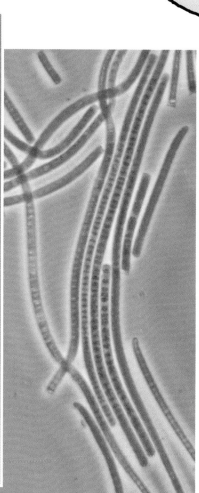

SPIRULINA

PHOTO: Courtesy of Carolina Biological Supply Company

READING SELECTION

cycling through the ecosystem by breaking down organic compounds, which releases the nutrients. One acre of soil may contain 6800 kilograms (1500 pounds) of bacteria.

The ability of bacteria to break down organic compounds has been exploited by facilities that process human waste. At wastewater treatment plants, bacteria are added to sewage; the bacteria eat the organic parts of the sewage, breaking them down in the process. Bacteria have even been used to help clean up oil spills by digesting some components of the petroleum.

So, don't think of bacteria as your enemies. They perform many useful jobs. What's more, they were one of the earliest life forms on Earth. Bacteria that can photosynthesize are believed to have been responsible for setting the stage for the rest of life to evolve. By photosynthesizing, they released oxygen, changing Earth's atmosphere into the oxygen-rich one that other organisms need. Once upon a time, bacteria ruled! Because they are so numerous and powerful, many people think they still do. ■

▶ THE SMALL KNOBBY SPHERES ON THE ROOT OF THIS PEA PLANT CONTAIN BACTERIA THAT FIX NITROGEN THE PLANT CAN USE.

PHOTO: Miloslav Kalab

DISCUSSION QUESTIONS

1. What would Earth be like without any bacteria?

2. In this reading selection, bacteria are referred to as "powerful." If you were to create a superhero Bacteria character, what powers would it have?

MIGHTY MICROBES

It's hard to believe that something visible only through a microscope could change the course of history. But it's true—and that something is microbes.

Microbes are microscopic organisms, so tiny that you need a microscope to see them—sometimes quite a powerful microscope. They live in the air, in water, and on just about any surface. Because organisms from several kingdoms fall into this category, microbes can be prokaryotes (organisms whose cells have no nuclei) or eukaryotes (organisms whose cells do have nuclei). Microbial prokaryotes include bacteria and archaebacteria; the eukaryotic microbes are a very diverse group of creatures. There are protists, like amoebae; there are fungi, microscopic algae, and even microscopic animals such as plankton and planaria. Much of life exists at the microbial level. So do viruses. Although viruses are not alive—they cannot reproduce on their own, and need to hijack living cells' reproductive machinery—many scientists classify them as microbes because they operate at a microbial level.

According to Dr. Leleng Isaacs, microbes have shaped the history of Earth and of human beings. Dr. Isaacs is a microbiologist—a scientist who studies tiny life forms. She has been studying microbes and their impact on history for many years. She and other scientists agree that humans are alive today because of microbes.

▶ DR. LELENG ISAACS TEACHES A LAB AT GOUCHER COLLEGE.

PHOTO: Courtesy of Goucher College

READING SELECTION
EXTENDING YOUR KNOWLEDGE

MICROBES: THE FIRST LIFE FORMS

When Earth was very young, the only life forms were microbes. Nevertheless, Dr. Isaacs says, being very small did not keep them from being very creative. For instance, scientists believe that one type of microbe, a blue-green alga, was the first to develop the ability to use sunlight and carbon to make food, or photosynthesize. As they photosynthesized, they made a waste product: oxygen.

According to Dr. Isaacs, in a way, this made microbes the first polluters. Oxygen is a pollutant because when oxygen is free inside a cell, it can be extremely destructive. But in this case, polluting was a good thing. The oxygen that was released into the atmosphere helped create the ozone layer. That layer protects Earth from dangerous radiation. Without it, new life forms could not have evolved. The oxygen buildup in the atmosphere also allowed the evolution of life forms that needed oxygen to survive—including humans.

BAD MICROBES, GOOD MICROBES

Some microbes cause diseases that can kill entire populations. We see this particularly when microbes are carried into a new area whose inhabitants have never had to deal with them before. For example, when Europeans first came to the Western Hemisphere, they brought diseases that had existed in Europe for a long time but which had not been present in the Americas. Some of those diseases, such as smallpox, infected the native peoples living in North and South America. Many natives died because they had no immunity to those diseases. Microbes helped to wipe out mighty civilizations such as the Incas.

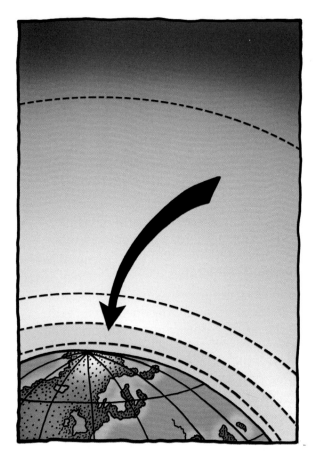

▶ THIS 15-KM (9.3-MI) LAYER OF OZONE, INDICATED BY THE ARROW, HELPS PROTECT US FROM THE HARMFUL RAYS OF THE SUN. (THE EARTH AND THE LAYERS OF ATMOSPHERE ARE NOT DRAWN TO SCALE.)

Today, we see similar threats from microbes. The human immunodeficiency virus (HIV), for instance, has infected hundreds of millions of people around the world. The disease it causes, acquired immune deficiency syndrome (AIDS), has spread fast. Scientists are working hard to find a vaccine that will prevent AIDS.

At the same time, microbes can do a lot of good. Many antibiotics are made from bacteria and other microbes. You probably have taken penicillin or amoxicillin for strep throat or some other kind of infection. Those medicines are made from mold, a type of fungus. Before the discovery and mass production of antibiotics in the middle of the 20th century, thousands of people died every year from diseases caused by bacteria.

TOO MUCH OF A GOOD THING

In TV ads, makers of soap and cleaning products brag that their products are "antibacterial." They say that's a good thing. Dr. Isaacs disagrees. She says overuse of antibacterial products and antibiotics is actually dangerous. Microbes can quickly become resistant to them. Then, scientists have to struggle to find newer, more effective medicines. Dr. Isaacs' advice? Use good old soap and water.

For Dr. Leleng Isaacs, microbiology is a fascinating field of study. Microbes are always changing and creating new challenges for scientists. She urges you to look into a microscope and discover the exciting world of microbes. ■

DISCUSSION QUESTIONS

1. Why were early microbes on Earth important to life forms that evolved later?

2. What would you say to convince a friend to buy a bar of regular soap instead of a liquid antibacterial soap?

EXPLORATION ACTIVITY: VERTEBRATES AND THEIR HABITATS

INTRODUCTION

Did you ever wonder how a snake can devour things more than twice its diameter? Or how a seal keeps from freezing in subzero temperatures? In this lesson, you will work with your group to explore what constitutes a habitat and to investigate a vertebrate—an animal with a backbone—such as that snake or seal. Your group then will divide into pairs. One pair will investigate how the vertebrate's body structure affects various aspects of its life, from what it eats and how it gets its food, to how it interacts with its own species and others. The other pair of students will research the vertebrate's habitat. Your group will work as a team to present your findings to the class. Your teacher will discuss the options for presenting your final project. You also will read a true story about an organism called *Daphnia* and its connections to Charles Darwin and his theory of evolution.

> **HOW LONG WOULD YOU LAST LYING ON THE ICE WITH NO CLOTHES FOR PROTECTION?**
>
> PHOTO: Courtesy of Carolina Biological Supply Company

> **THIS BAT-EATING SNAKE IS CAPABLE OF SWALLOWING CREATURES TWICE ITS DIAMETER.**
>
> PHOTO: Courtesy of Carolina Biological Supply Company

OBJECTIVES FOR THIS LESSON

Select a vertebrate and research how the structure of its body parts influences the way those parts function.

Research your vertebrate's habitat to discover the biotic and abiotic factors that might affect its ability to survive.

Share your findings with the class using a presentation method approved by your teacher.

Read about an organism called *Daphnia* and decide whether its rapid evolution supports Charles Darwin's ideas.

MATERIALS FOR LESSON 7

For you

1 copy of Student Sheet 7.1: Exploration Activity Schedule

1 copy of Student Sheet 7.2: Exploration Activity Scoring Rubrics

For your group

1 copy of Inquiry Master 7.1: Lists of Vertebrates

GETTING STARTED

1 Read silently as a classmate reads aloud the Introduction to this lesson.

2 Watch the DVD *Body by Design: Form and Function* and take notes based on your teacher's instructions.

3 Brainstorm with your group a list of vertebrate body parts with unique adaptations for performing one or more functions.

4 Share your list with the class.

▶ **A BLACK-FOOTED FERRET USES ITS SHARP CLAWS TO DIG BURROWS.**

PHOTO: Ryan Hagerty/U.S. Fish and Wildlife Service

PART 1

INTRODUCING THE EXPLORATION ACTIVITY

PROCEDURE

1 Listen while your teacher introduces the Exploration Activity, including appropriate research methods, reference requirements, deadlines, and presentation formats. Follow along as your teacher reviews Student Sheet 7.1: Exploration Activity Schedule and Student Sheet 7.2: Exploration Activity Scoring Rubrics. Your teacher will tell you the appropriate point values for each part of the scoring rubric.

2 Read "Habitats as Homes" on pages 109-113 to find out more about habitats in general.

PART 2

CHOOSING A VERTEBRATE

PROCEDURE

1 Work with your group to choose a vertebrate to research. Refer to a variety of sources including the lists provided by your teacher, your school media center, and your own resources.

2 Obtain your teacher's approval for your group's final choice.

3 Decide which pair from your group will research how your vertebrate's various body parts are suited to their functions and which pair will research the vertebrate's habitat.

4 Read "Animals with Backbones" on pages 100-101.

READING SELECTION

BUILDING YOUR UNDERSTANDING

ANIMALS WITH BACKBONES

A vertebrate is an organism with a backbone. Vertebrates are grouped within the phylum Chordata of the Animal kingdom and are further separated into five major classes:

Class Mammalia—Warm-blooded animals with hair or fur and mammary glands

▶ THIS BROWN BEAR, WITH ITS THICK COAT OF FUR, IS A RUGGED SPECIMEN OF A MAMMAL.

PHOTO: Courtesy of Carolina Biological Supply Company

Class Aves—Warm-blooded animals with feathers and hollow bones

▶ THESE WAXED ALBATROSSES ARE WARM-BLOODED, WHICH MEANS THAT THEIR BODIES MAINTAIN A CONSTANT INTERNAL BODY TEMPERATURE. WHAT DO YOU THINK "COLD-BLOODED" MEANS?

PHOTO: Courtesy of Carolina Biological Supply Company

Class Reptilia—Cold-blooded animals with scales that lay their eggs on land

▶ THIS SNAKE HAS A THREATENING RATTLE THAT WARDS OFF ENEMIES.

PHOTO: Courtesy of Carolina Biological Supply Company

Class Pisces—Cold-blooded animals that live in water (they are often divided into smaller groups, such as jawless fish, cartilage fish, and bony fish)

▶ THIS GREEN SUNFISH IS JUST ONE OF MANY SPECIES OF SUNFISH YOU MIGHT FIND IN A POND OR LAKE.

PHOTO: Courtesy of Carolina Biological Supply Company

Class Amphibia—Cold-blooded animals that live part of their life in water and part on land (they breathe by means of gills when young, but develop lungs as adults); they lay their eggs in water ■

▶ THIS LEOPARD FROG IS WELL CAMOUFLAGED AMIDST THE DUCKWEED. THE POSITION OF ITS EYES ALLOWS IT TO OBSERVE ITS SURROUNDINGS WITHOUT EXPOSING ITS WHOLE BODY TO PREDATORS.

PHOTO: Courtesy of Carolina Biological Supply Company

PART 3
GATHERING DATA
PROCEDURE

1 Listen while your teacher summarizes the steps that each pair of students must follow. Review Figure 7.1.

2 The pair that is researching body parts should take the following steps:

A. Discuss the DVD *Body by Design: Form and Function*, which featured the numerous ways in which body structures are adapted to perform their functions.

B. Conduct research using a variety of references, including books, encyclopedias, DVDs, and the Internet. Remember to research both internal and external structures. Your final report must include the scientific name of your vertebrate, and at least five body structures, and how these structures are suited for performing their functions.

3 The pair researching your vertebrate's habitat should take the following steps:

A. Include in your research those factors caused or produced by living beings (biotic) and those factors not caused or produced by living beings (abiotic). Examples of biotic factors include the following:

- predators, or lack of them
- food preferences and supply
- population density
- other species with which your vertebrate interacts in its habitat
- disease

Examples of abiotic factors include the following:

- temperature range
- precipitation
- water
- topography
- wind
- shelter
- nesting site and a place to raise young
- escape routes from predators

B. Also include information on the impact of your vertebrate and its habitat on humans and vice versa. Include drawings, photos, or computer-generated images of your vertebrate's habitat, preferably with an image of your vertebrate included.

C. Conduct research using a variety of references, including books, encyclopedias, DVDs, and the Internet.

Homo sapiens

Hair—Provides insulation and protection from injury. May vary in color, diameter, texture, and surface area coverage. The degree of insulation and protection is dependent partially on these variables.

Brain—Large, well-developed brain; capable of advanced reasoning and decision-making, which contributes to the human's ability to survive, change, and develop complex behavior patterns.

Outer ear—Outer funnel that collects sound and directs it inward to the middle and inner ear. Only mammals have this funnel-shaped outer ear. Scientists are unclear whether it helps humans hear better.

Teeth

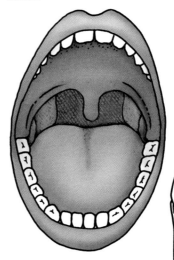

Back teeth are flat for chewing.

Front teeth are sharp and pointed for tearing. These features allow humans to eat a variety of foods.

Joint—Place where bones meet. Bones are connected by elastic ligaments that permit flexibility and movement.

▶ **A VERTEBRATE IN ACTION**
FIGURE **7.1**

PHOTO: National Science Resources Center

PART 4

PRESENTING YOUR RESEARCH PROJECT

PROCEDURE

 Your group will share its research project with the class. Listen while your teacher reviews various product and presentation options.

2 Your pair may choose any of the presentation methods approved by your teacher. Each pair in your group may choose different methods. Whatever method you choose, remember that the pair researching body parts must include in its final product all of the elements listed in Part 3 Step 2B, and the pair researching habitats must include all of the elements listed in Part 3 Steps 3A and 3B.

CHARLES DARWIN AND HIS THEORY OF EVOLUTION

In 1859, British naturalist Charles Darwin (1809–1882) published his groundbreaking work, *Origin of Species*. In this remarkable book, Darwin explained his theory of evolution, which was based on many years of observations. The book illustrated several important points:

- Many variations exist within species. For example, some humans are taller than others; some giraffes have longer necks than others.
- All organisms compete for the resources they need to survive.
- Organisms can produce more offspring than can survive given the quantity of resources available.
- Organisms that are fit and better able to deal with changes in their environment tend to survive and reproduce, passing their desirable traits on to their offspring. This is known as "natural selection," or "survival of the fittest."

▶ CHARLES DARWIN
PHOTO: Courtesy of the National Library of Medicine

3 If your pair decides to create a poster, for example, you would take the following steps:

A. Prepare a draft of your poster on 8½" x 11" paper for approval by your teacher. Include your research sources on a separate piece of paper.

B. After receiving your teacher's approval, transfer your information to a 24" x 36" poster. Figure 7.1 illustrates one way of documenting research findings in the form of a poster. Several, but certainly not all, of this familiar vertebrate's structures are labeled, with a brief explanation of their functions.

4 Your teacher will assign a due date for the draft of your team's product. (If you choose a presentation method other than a poster, ask your teacher how to provide a draft of your research for approval.)

5 Arrange time for you and your partner to practice your presentation. Use the scoring rubric on Student Sheet 7.2 as a guide for assessing your project. Your teacher will explain the point values for each step of each rubric.

REFLECTING
ON WHAT
YOU'VE DONE

1 Read the passage about Charles Darwin on page 104 and the article entitled "*Daphnia*'s Change of Appetite" on pages 106–108. Then write a minimum of 150 words explaining which of Darwin's ideas are addressed in the reading selection and how those ideas are supported. Turn in your work to your teacher by the date requested in the schedule provided on Student Sheet 7.1.

Daphnia's Change of Appetite

Have you ever thrown away something you were about to eat because it did not smell right? Or dumped milk down the drain because it seemed spoiled? If so, you probably did the right thing. Eating or drinking food that has spoiled can make you sick.

But what if all your food turned bad and you had nothing else to eat? If your body would not accept the bad food, you would starve. In the

▶ POLLUTION OFTEN LIMITS THE TYPES OF ORGANISMS THAT ARE ABLE TO SURVIVE IN A LAKE OR POND.

PHOTO: National Science Resources Center

natural world, many creatures face this problem. This is especially true of organisms that live in lakes close to cities and towns, where pollution often is a problem.

Pollution from human activity has gradually poisoned the food of many aquatic, or water-dwelling, organisms. Because these organisms live within confined bodies of water, they can't leave their homes to seek cleaner waters. If none of the organisms is able to survive the pollution, the entire species will perish in that habitat.

WHAT'S FOR DINNER?

In Lake Constance in Germany, one tiny creature has defied the odds. It is called *Daphnia*, and it belongs to a biological class called Crustacea. The best-known crustaceans are shrimp, crabs, and lobsters. *Daphnia* is much smaller than any of these. You might call it a "shrimp's shrimp."

Since 1970, human-caused pollution in Lake Constance has killed off much of *Daphnia*'s one-time favorite food—harmless green algae. While the green algae could not survive the pollution, a different species, a poisonous blue-green form of algae called cyanobacteria, flourished. It now dominates the lake. Cyanobacteria is not only dangerous to *Daphnia*; humans also can get very sick if they drink water containing these algae.

After the harmless green algae were gone, *Daphnia* in Lake Constance turned to the other available source of food—the dangerous cyanobacteria. Surprisingly, the entire population of *Daphnia* did not die off. Those rare *Daphnia* endowed with genes that allowed them to tolerate the new diet of cyanobacteria survived, reproduced, and passed on their tolerance to their offspring. If there had been no *Daphnia* with this tolerance in Lake Constance, their species in the lake would have perished.

DAPHNIA FROM OLD EGGS

How do scientists know that *Daphnia* have evolved? Through an amazing bit of detective work.

Scientists knew that each year, after *Daphnia* laid their eggs, some of the eggs became buried at the bottom of Lake Constance. Year after year, the sediment covered more and more eggs. So the deeper the eggs are buried, the older they are.

A group of scientists decided to dig up some eggs that had been buried about 30 years ago, before the lake became polluted, as well as some eggs that were buried about 20 years ago. Then the scientists brought the eggs to the laboratory, where, incredibly, the 30-year-old eggs were still able to hatch! They raised the newly hatched *Daphnia* in the laboratory, feeding deadly cyanobacteria to both the pre- and post-pollution creatures.

▶ *DAPHNIA*

PHOTO: (2205) Are We Underestimating Species Extinction Risk? PLos Biol 3(7): e253. doi: 10.1371/journal. pbio.0030253/Creative Commons Attribution 2.5 License

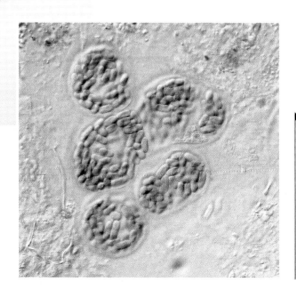

▶ THIS HARMLESS-LOOKING ORGANISM, A TYPE OF CYANOBACTERIUM, CAN BE VERY HARMFUL TO MOST *DAPHNIA*.

PHOTO: NASA Ames Research Center

What did the scientists discover? *Daphnia* from the older eggs couldn't eat cyanobacteria and survive. Their diet evidently had been the green algae, and they couldn't "stomach" the blue-green stuff. And—you guessed it—*Daphnia* hatched from 20-year-old eggs ate the toxic blue-green algae without a problem.

The scientists were surprised. *Daphnia* had evolved in only about 10 years. In the process, the species had survived in the lake, and, by consuming the poisonous cyanobacteria, had helped make Lake Constance safe for humans again. ■

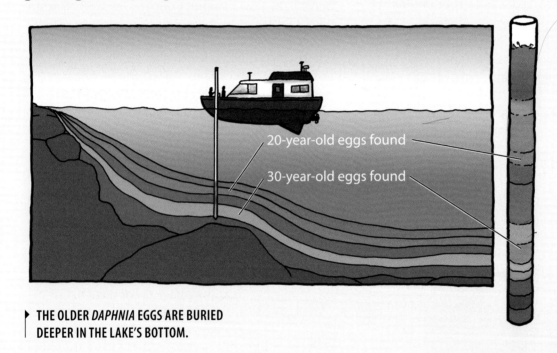

20-year-old eggs found

30-year-old eggs found

▶ **THE OLDER *DAPHNIA* EGGS ARE BURIED DEEPER IN THE LAKE'S BOTTOM.**

DISCUSSION QUESTIONS

1. How long did it take *Daphnia* to adapt to its new diet? Explain how this adaptation occurred.

2. Would a larger, longer-lived animal like a turtle likely have adapted more quickly? Or more slowly? Explain your reasoning.

HABITATS
AS HOMES

A habitat is any place where a plant or animal lives. A habitat can be a desert for a cactus or a rattlesnake, an ocean for a whale or a seaweed, a grassland for a bison or a butterfly, or a rainforest for a monkey or a rubber tree. For some parasitic worms, a habitat can be an animal's intestine. For some species of fungi, a habitat can be a single tree.

In other words, a habitat is home. It's the place that supplies an organism with all it needs—food, water, shelter, and a place to bear and raise its young. What is your habitat like?

▶ **THIS FUNGUS LIVES VERY COMFORTABLY ON AND WITHIN THIS TREE.**

PHOTO: Jim Kuhn/creativecommons.org

READING SELECTION
EXTENDING YOUR KNOWLEDGE

People's habitats change throughout their lives. A city child may grow up to be a biologist, and make a home in a rainforest or desert, near the life forms she studies. So too do other animals change their habitats at certain times in their lives. Consider the sea turtle, for example. Females of the world's seven sea turtle species all come to shore to nest and lay their eggs. When the babies hatch, they rush to the sea, which becomes their habitat. But for a very brief time in the female sea turtle's life, the ocean shore is her habitat. Male sea turtles never need to return to land. So for adult males, home is found only in the ocean.

Within the vast ocean, though, sea turtles find different types of habitats. Some leatherback sea turtles spend part of the year in Alaskan waters, feeding on jellyfish. Hawksbill turtles feed along tropical coral reefs. Young Kemp's Ridley sea turtles mature in the reefs of the Sargasso Sea.

▶ OF ALL THE HATCHLINGS THAT NOW CALL THE OCEAN THEIR HOME, ONLY THE FEMALES EVENTUALLY WILL RETURN TO LAND TO LAY THEIR EGGS.

PHOTO: Courtesy of Carolina Biological Supply Company

▶ **THESE SALMON STRUGGLE TO SWIM UPSTREAM TO THEIR SPAWNING GROUNDS.**

PHOTO: Courtesy of Carolina Biological Supply Company

Salmon alternate between two different habitats. Like the female sea turtle, they must have both for their species to survive. Salmon hatch in freshwater streams and migrate to the sea, where they live for a few years. Then they swim back up the streams of their birth to lay eggs and die.

Whatever the habitat, the organisms that live there—especially humans—have an impact on it. People chop down forests to clear room for fields and homes, and to harvest the timber. We use the local water and soil; we take up room that was once used by other organisms. All creatures change their habitats.

And, slowly but surely, habitats also change their creatures. This is evolution. For example, when a group of rabbits moves to a new habitat, some of them will survive and reproduce better than others. If rabbits move to a new habitat with short grasses, where they're easily seen from overhead, the rabbits with the sharpest eyes will be safest from the hawks that prey on them. They'll survive to breed, and their offspring's ability to spot hawks will be unusually good, too. Over time, the population of rabbits living in that habitat will have eyes that are sharper, on the whole, than the eyes of the rabbits in the old habitat. They will have evolved.

READING SELECTION
EXTENDING YOUR KNOWLEDGE

It's important to understand that individual organisms do not adapt and evolve. Whole populations adapt and evolve. Each time a new litter of rabbits is born, some of them will have sharper eyes than the others. The rabbits' eyes won't change while they're alive. But if you step back and look at many generations of rabbit litters in the habitat, you'll see that in the last generation, there are more sharp-eyed rabbits than there were in the first generation.

Evolution is not a decision, and it doesn't happen on purpose. Rather, the organisms whose traits give them an edge in the habitat—for instance, the ones that are faster, or better camouflaged, or better able to climb trees for food—tend to have more offspring. These useful traits are passed on to their offspring, who then have more offspring, and so on. This is called "natural selection," because nature automatically selects the organisms that reproduce best in their habitats.

When a population changes to do better in a habitat, we say that it adapts. Adaptation is a continuous process, because not only do populations move, but habitats change. Think back to the sharp-eyed rabbits. The rabbits may find that the local climate is suddenly drier than it used to be, or that another grass-eating species has moved in and is competing for food. Over time, the grassland itself may become a forest. Sharp eyes will not be enough to keep the rabbit colonies thriving; new adaptations will be necessary.

Many plants and animals evolve in ways that leave them able to survive only in specific habitats. Australia's koala bear, for example, feeds only on the leaves of a few species of eucalyptus trees, making it very specialized. Species like the koala are vulnerable to changes in their habitat. If disease were to kill off Australia's eucalyptus trees, for example, the koala in that area would soon perish. By contrast, for their survival, cabbage white

▶ THIS KOALA IS RESTING COMFORTABLY ON THE BRANCHES OF A EUCALYPTUS TREE, THE LEAVES OF WHICH ARE ITS MAIN DIET.

PHOTO: emmett anderson/creativecommons.org

butterflies depend on plants in the mustard family, which include more than 3,000 species in over 300 genera. Therefore, the cabbage white may inhabit the many places in the world in which varieties of mustard plants are found.

Vertebrates in two classes, mammals and birds, are warm-blooded. This means that they maintain a constant body temperature. The other three classes of vertebrates—reptiles, amphibians, and fish—consist of animals that are cold-blooded, meaning that they take on the temperature of their surroundings. Because they can maintain a constant body temperature, mammals and birds are generally able to adapt to a wider range of temperatures and thus a wider range of habitats.

If a population does not adapt to its changing environment, one of two things will happen. It will move to a place where it can live more easily, or it will shrink. It may even go extinct. Finding a new, suitable habitat isn't always easy. In fact, scientists estimate that most species of organisms that have ever lived are now extinct.

Some organisms, such as certain species of mammals, birds, and fish, migrate to other areas during certain times of the year. They might migrate to follow food supplies, seek more suitable temperatures, find places to reproduce, or for a combination of reasons. As a result, their habitats temporarily change.

Other organisms have ways to avoid stressful changes in their environment. Seeds may go into periods of rest, or dormancy. The cabbage white butterfly larva forms a protective chrysalis, a stage of metamorphosis in which it may exist until conditions are favorable for it to emerge as an adult.

A habitat may be as tiny as a drop of water or a pinch of soil, or as large as an ocean or a forest, as long as it provides an organism those things it needs for its survival—food, shelter, and a place to reproduce. They are always changing, and so are the organisms within them. ■

DISCUSSION QUESTIONS

1. Kaitlin seems to be at the pool most of the time during the summer. Is the pool a habitat? Why or why not?

2. What sorts of change in a habitat could force a population to evolve or go extinct?

REVISITING YOUR POND

INTRODUCTION

In Lesson 4, you observed and sketched the layers of your pond and observed its water through a microscope. In this lesson, you will take another look at your pond to see what changes, if any, have occurred. You also will read "The Changing Pond" at the end of this lesson to find out how ponds develop and change over time in a natural setting.

▶ **WHAT STAGE IN THE LIFE OF A POND DOES THIS PHOTO REPRESENT?**

PHOTO: National Science Resources Center

OBJECTIVES FOR THIS LESSON

▸ Observe your pond, then sketch it and label the living and nonliving things you included in your sketch.

▸ Look for evidence of succession in your pond.

▸ Use a compound microscope to observe, draw, and identify microorganisms from different depths in your pond.

▸ Determine the average daily increase in the number of *Lemna* fronds over three weeks.

▸ **MATERIALS FOR LESSON 8**

For you

Your copy of Student Sheet 2.3a: Guidelines for Scientific Drawings

1 copy of Student Sheet 8.1a: My Pond and Its Organisms

1 copy of Student Sheet 8.2: Average Daily Increase in the Number of *Lemna* Fronds

Your copy of Student Sheet 4.2: Sketches of Pond—Macro and Micro

For your group

1 copy of Student Sheet 8.1b: Common Freshwater Microorganisms

1 pond (from Lesson 4)

2 compound light microscopes

2 depression slides

2 hand lenses

2 coverslips

2 plastic pipettes

1 small cotton ball

1 black marker

2 metric rulers, 30 cm (12 in.)

GETTING STARTED

1 Do not shake or stir your pond.

2 With your group, take turns using the hand lenses to observe your pond at eye level. Make sure there is enough light to see into it clearly.

3 Compare your pond with the one you drew on Student Sheet 4.2. In your science notebook, list two ways in which your pond has changed and two ways in which it has remained the same. 🖉

4 Share your observations with the class.

▶ **WHAT DO YOU THINK THIS BULLFROG SEES IN THE POND?**

PHOTO: John J. Mosesso/life.nbii.gov

OBSERVING AND DRAWING MY POND AND ITS MICROBES

PROCEDURE

1 Sketch your pond in the box at the top of Student Sheet 8.1a: My Pond and Its Organisms. Include everything you can see, such as distinct layers and any organisms visible through a hand lens. Label the nonliving and living things.

2 Add a few strands of cotton to the depression on your slide.

3 Use a pipette to obtain water from an area near the surface of your pond. Add one drop to the cotton on the slide. Place a coverslip on the slide.

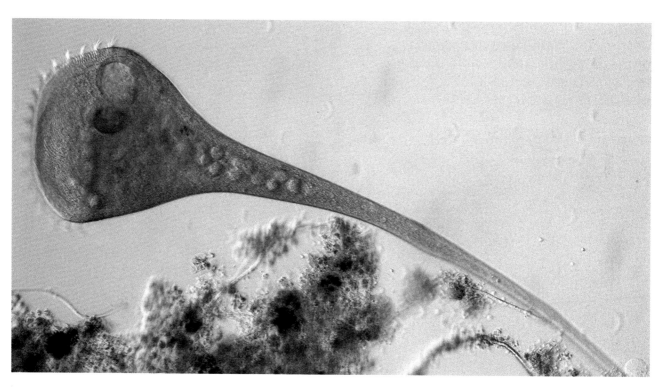

▶ *STENTOR* IS A LARGE PROTIST THAT ATTACHES BY ITS BOTTOM TO DEBRIS OR AQUATIC PLANTS. THE HAIR-LIKE CILIA SURROUNDING ITS MOUTH MOVE LIKE TINY OARS, SWEEPING IN SMALLER PROTISTS AND BACTERIA THAT SWIM NEARBY. YOU MAY FIND ONE OR MORE OF THESE IN YOUR POND.
FIGURE **8.1**

PHOTO: Courtesy of Carolina Biological Supply Company

Inquiry 8.1 *continued*

4 Focus on the drop of water under a magnification of 100x. While looking through the eyepiece, slowly move the slide around to locate microorganisms. Be sure to look for algae and other vegetation as well as mobile microbes.

5 Sketch as many of the organisms as you can in the circles provided on Student Sheet 8.1a. Follow the guidelines for scientific drawings on Student Sheet 2.3a. Change magnifications as needed, using whichever magnification gives the best view of each organism.

▸ *VORTICELLA* ARE PROTISTS THAT CAN LIVE INDEPENDENTLY BUT ARE USUALLY FOUND IN CLUSTERS. THEY ATTACH TO AQUATIC PLANTS USING A STALK THAT CONTRACTS LIKE A SPRING WHEN DISTURBED. CAN YOU FIND ONE IN YOUR POND?
FIGURE **8.2**

PHOTO: Courtesy of Carolina Biological Supply Company

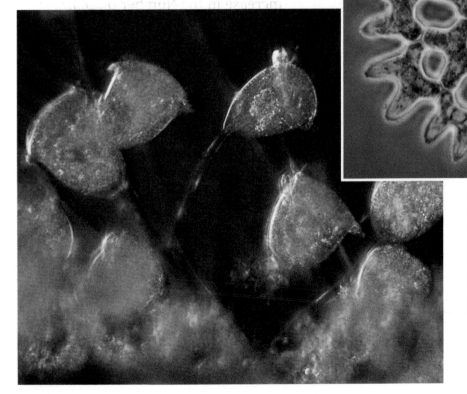

▸ *PEDIASTRUM* ARE GREEN PROTISTS THAT LIVE IN COLONIES OF 32–64 CELLS. THEY CAN USUALLY BE FOUND AT THE EDGE OF A POND, BELOW THE TOP LAYER OF SAND AND SOIL.
FIGURE **8.3**

PHOTO: Courtesy of Carolina Biological Supply Company

6. Repeat Steps 4 and 5 with water samples taken carefully from the bottom level of your pond. On Student Sheet 8.1a, draw and label each new organism that you observe. Obtain another copy of Student Sheet 8.1a if you need more than four circles. Use Student Sheet 8.1b: Common Freshwater Microorganisms and Figures 8.1–8.3 to help you identify unfamiliar microorganisms.

7. Share your slides with other students, especially if you find something particularly interesting.

DETERMINING THE AVERAGE DAILY INCREASE IN THE NUMBER OF *LEMNA* FRONDS

PROCEDURE

1. On the day you created your pond, you placed *Lemna* plants in it, counted the total number of fronds on those plants, and recorded that figure in your science notebook. Count the number of fronds you now have in your pond. Record both figures in the appropriate boxes on Student Sheet 8.2: Average Daily Increase in the Number of *Lemna* Fronds.

2. Using those figures, calculate and record the average increase per day in the number of *Lemna* fronds.

3. Share your data with the class.

Although you are investigating asexual reproduction of *Lemna*, these tiny plants also reproduce sexually. Relatives of *Lemna*, also in the duckweed family, are *Wolffia* species, which produce the world's smallest fruits and flowers.

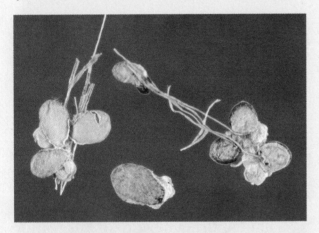

▶ TINY LEAVES AND FLOWERS OF SEXUALLY REPRODUCING *LEMNA GIBBA*
FIGURE **8.4**

PHOTO: Jose Hernandez @ USDA-NRCS PLANTS Database

▶ EVEN TINIER DUCKWEED, A *WOLFFIA* SPECIES
FIGURE **8.5**

PHOTO: Christian Fischer/creativecommons.org

REFLECTING
ON WHAT
YOU'VE DONE

1 Based on what you have learned in the inquiries and readings in this lesson, respond to the following in your science notebook:

A. You have observed your pond, past and present, with and without magnification. Describe the changes you have observed.

B. Are the new organisms you observed during this lesson really new to your pond environment? Where do you think they came from?

C. Why do you think your teacher asked you to add rice grains to your pond? What do you think happened to them?

D. Do the changes occurring in your pond give you a clue as to why real ponds eventually fill in? Explain.

E. Look at the photo on page 114. What stage in the life cycle of a pond do you think the photo represents? Use a number from the illustrations on page 122 to represent the stage. Defend your answer.

F. What factors do you think may have influenced the average daily increase in the number of *Lemna* fronds?

G. If *Lemna* reproduced to cover the entire top of your pond—or a real pond—what effects do you think this might have on the pond and its inhabitants?

The CHANGING POND

PHOTO: National Science Resources Center

Think about the day you created your pond; think about the substances that you added to the cup. They included gravel, soil, decaying leaves, straw, rice grains, water, and *Lemna*. When you observed water from your pond through a microscope on that day, you probably saw little evidence of living organisms.

Now that your pond is older, it looks quite different. All ponds develop and change. It's true that natural ponds don't begin their existence the way your pond did; natural ponds begin by accident, as shallow depressions where water has collected. They are usually no more than 3-4 meters deep, and sunlight can reach their bottoms. Even so, they go through changes, just as the pond in your cup has done.

READING SELECTION
EXTENDING YOUR KNOWLEDGE

▶ SUCCESSION OF A POND

As they age, natural ponds go through a gradual process of change known as succession. First, a depression fills with water, deep enough to seem like a semi-permanent feature of the landscape. The water attracts nearby wildlife. Land animals and birds that come to inhabit the area bring in plant seeds and spores. Seeds and spores also may blow or drift in. Over time, plants grow from the seeds, and vegetation in and around the pond increases. Some types of plants float on the surface; others grow on the bottom. Still other types of vegetation grow on the banks of the pond.

In addition, wind and rain can carry decaying leaves, straw, and soil into ponds. (In your pond, you added the straw, leaves, and soil.) Microbes often dwell on these substances, and are carried with them into the pond. Many of these microbes—particularly protists—have lain dormant in an envelope, called a cyst, which surrounds them to protect against unfavorable environmental conditions such as cold or hot temperatures or dryness. When they are carried into ponds and exposed to water and suitable temperatures, the microbes are revived. The cysts wear away, and the microbes resume their normal life activities.

Over many years, as each type of vegetation in a pond dies and decomposes, plant debris accumulates at the pond's bottom. Animals die and their remains also fall to the bottom of the pond. The mixture of materials is called "detritus." The dead plant and animal materials provide nourishment for organisms called decomposers—

bacteria, fungi, certain types of insects, and worms. These organisms thrive on organic matter. Eventually, over many years, the products of decomposition fill up the pond. The "bottom" of the pond literally rises close to the top.

At this point, plants can root in the decomposed matter at the bottom and send their leaves into the air. The pond becomes a marsh or swamp. As the area continues to fill in, trees begin to grow in the water. Over a long time, as much as hundreds of years, the pond fills in and slowly dries out, becoming either a forest or a grassland, depending on the local climate.

With this, the pond appears to have completed its succession. It may remain a stable community in which the numbers and types of organisms are in relative balance. More likely, other disturbances will occur, generating more change. ■

DISCUSSION QUESTIONS

1. Do all ponds eventually fill in? Why or why not?

2. What sorts of natural events (disturbances) might start the formation of new ponds?

A POND'S HIDDEN LIFE

Life in a pond is not always obvious. The most important sources of food for all organisms in a pond are microbes. In addition to being food sources, many microbes, such as algae, release oxygen needed by other organisms in the water. Larger, floating, rooted, and submerged plants also contribute to the food supply.

There are many food chains and food webs in a pond ecosystem. Microscopic meat eaters (carnivores) feed on microscopic plant eaters (herbivores), which feed on algae and other types of tiny, plant-like organisms. Predatory insects in and around a pond feed on microscopic carnivores. These insects are eaten by larger insects and spiders, which in turn may be eaten by birds, frogs, fish, and other animals. These organisms will have been attracted to the pond by the food and shelter this habitat offers.

Look again at your pond. What do you think it will look like after more time passes?

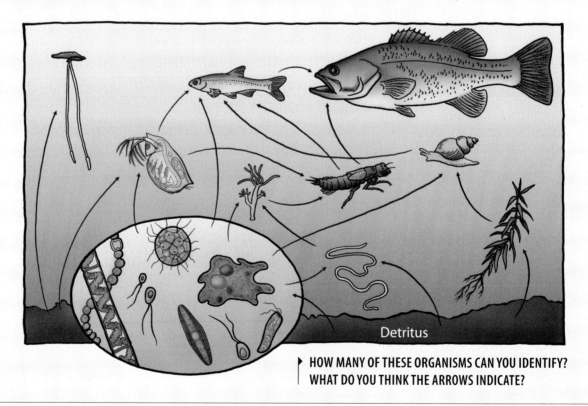

Detritus

▶ HOW MANY OF THESE ORGANISMS CAN YOU IDENTIFY?
WHAT DO YOU THINK THE ARROWS INDICATE?

INTRODUCING *DAPHNIA*

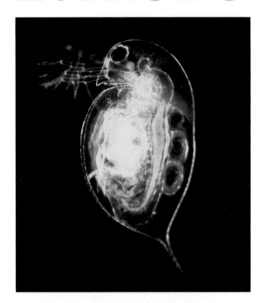

▶ *DAPHNIA*'S TRANSPARENT OUTER SHELL
MAKES IT EASY TO OBSERVE ITS INTERNAL
STRUCTURES.

PHOTO: Paul Hebert/Gewin V (2005) Functional Genomics
Thickens the Biological Plot. PLoS Biol 3(6): e219. doi:10.1371/
journal.pbio.003029/Creative Commons Attribution 2.5 License

INTRODUCTION

In this lesson, you will explore some features and behaviors of an organism called *Daphnia*. You will observe the organism through a microscope and prepare a scientific drawing of a *Daphnia* and some of its structures. Finally, you will measure a *Daphnia*'s heart rate before and after the organism has been treated with two different chemicals.

OBJECTIVES FOR THIS LESSON

▶ Observe, sketch, and measure a *Daphnia* and identify its major structures.

▶ Determine the heart rate of a *Daphnia* under various conditions.

▶ Update your organism photo card for *Daphnia*.

▶ MATERIALS FOR LESSON 9

For you

Your copy of Student Sheet 2.3a: Guidelines for Scientific Drawings

1   copy of Student Sheet 9.1: Template for *Daphnia* Drawing

1   copy of Student Sheet 9.2a: Table for Recording Heartbeats of *Daphnia*

1   copy of Student Sheet 9.2b: Effect of Alcohol and Cola Solutions on the Heart Rate of *Daphnia*

For your group

1   set of organism photo cards
2   compound light microscopes
2   depression slides
2   strands of cotton
2   coverslips
4   *Daphnia* (2 per class period)
2   metric rulers, 30 cm (12 in.)
2   transparent rulers
4   pencils or fineline markers

GETTING STARTED

1 *Daphnia* are related to lobsters and shrimp. List in your science notebook some of the characteristics you think these three organisms share. Use Figures 9.1 and 9.2 and the Introduction photo on page 124 for reference.

2 Record your answers to the following questions. Be ready to share them with the class.

A. What kind of skeleton do *Daphnia*, lobsters, and shrimp have?

B. What are the advantages and disadvantages of this kind of skeleton over the type of skeletons humans have?

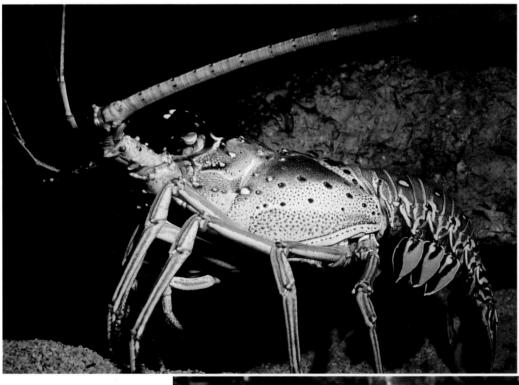

▶ **SPINY LOBSTER**
FIGURE **9.1**

PHOTO: Courtesy of Carolina
Biological Supply Company

▶ **SHRIMP**
FIGURE **9.2**

PHOTO: Courtesy of Carolina
Biological Supply Company

PREPARING A SCIENTIFIC DRAWING OF A *DAPHNIA*

PROCEDURE

1 You will work in pairs for this inquiry. Have one student from your pair go to the materials station to obtain a slide with a *Daphnia* in a drop or two of spring water. One or two strands of cotton under the coverslip will have isolated the *Daphnia* in a small area for viewing. Do not press down on the coverslip.

2 Use your microscope to focus on the *Daphnia* under the highest power at which you can see the entire organism.

3 Draw the *Daphnia* in as much detail as you can on Student Sheet 9.1: Template for *Daphnia* Drawing. Follow the guidelines for scientific drawings listed on Student Sheet 2.3a. Title your drawing "*Daphnia*: The Water Flea." Refer to "The Transparent Water Flea" on pages 132–133 for information about how to label your drawing.

4 Use your transparent ruler to measure the length of the *Daphnia*. Record the length, following the guidelines for scientific drawings.

Inquiry 9.1 continued

5 Take the following steps to observe and identify specific structures to include in your drawing. (Refer to Figure 9.3 and the illustration in "The Transparent Water Flea.")

A. Look for the intestine, which runs from the mouth to the anus. Notice its color. Discuss with your partner why you think it is this color.

B. Locate the *Daphnia*'s heart. Notice how rapidly it beats.

C. Find the "brood chamber," a sac located just below the heart of the female *Daphnia*. Discuss with your partner what might be found inside a brood chamber.

D. Focus on your *Daphnia*'s eye. Switch the microscope to a higher power to observe the eye in greater detail. Discuss with your partner how its structure differs from that of a human eye.

E. Focus on one of the antennae under high power. Discuss with your partner one possible function of the antennae.

F. Focus on one of the legs under high power. Notice the bristles. Discuss with your partner what you think the function of the bristles may be.

6 When you and your partner have completed your drawings, move to Inquiry 9.2 if instructed to do so by your teacher. Use the same slide but get a fresh *Daphnia* for this inquiry if you will be completing the inquiry during the same period. Follow your teacher's instructions for returning the *Daphnia* to the culture container.

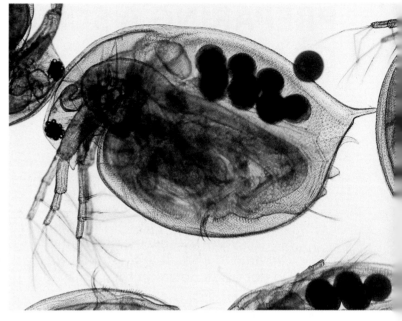

▶ YOU CAN EASILY SEE THE DIGESTIVE TUBE, HEART, AND BROOD CHAMBER IN THE *DAPHNIA* AT THE CENTER OF THIS PHOTO.
FIGURE **9.3**

PHOTO: Courtesy of Carolina Biological Supply Company

EXPLORING THE EFFECT OF ALCOHOL AND COLA SOLUTIONS ON THE HEART RATE OF *DAPHNIA*

PROCEDURE

1 Read all of the Procedure Steps carefully. They will give you the information necessary to prepare a data table in the box provided on Student Sheet 9.2b: Effect of Alcohol and Cola Solutions on the Heart Rate of *Daphnia*. You will record your data and your partner's data on this table.

2 Observe the *Daphnia* closely. Practice measuring its heartbeat using the following technique:

A. Place a pencil or fineline marker in your writing hand and hold the tip just above the middle of the top left box on Student Sheet 9.2a: Table for Recording Heartbeats of *Daphnia*.

B. For 10 seconds, tap the point of your pencil in the first box, making a mark each time the *Daphnia*'s heart beats as shown in Figure 9.4. Have your partner time you and call "Stop" when 10 seconds are up. If you need more practice, make your marks in the right box in the top row.

▶ EACH DOT REPRESENTS ONE BEAT OF THE *DAPHNIA*'S HEART.
FIGURE **9.4**

Inquiry 9.2 continued

3 When you have finished practicing, move your pencil so that you are holding the tip above the middle of the first box in the second row. Have your partner time you for 10 seconds while you mark each heartbeat.

4 Count the number of marks to determine the number of times your *Daphnia*'s heart beats in 10 seconds. Decide with your partner how to use this figure to calculate the *Daphnia*'s heart rate per minute. Record this calculation in the appropriate place on your data table.

5 Repeat Steps 3 and 4, using a different box for your marks. Use your data from the two trials to calculate the average heart rate of the *Daphnia* in spring water.

6 Switch roles with your partner and repeat Steps 2–5.

7 Watch and listen as your teacher demonstrates how to add an alcohol or cola solution to the *Daphnia* slides.

8 One pair in your group will measure a *Daphnia*'s heart rate while the organism is immersed in a weak alcohol solution. The other pair will do the same while its *Daphnia* is immersed in a weak cola solution. Before you begin, write one-sentence answers to the following questions on Student Sheet 9.2b:

A. How do you think alcohol will affect the heart rate of *Daphnia*? Why?

B. How do you think cola will affect the heart rate of *Daphnia*? Why?

9 Have one pair in your group add a drop of weak alcohol solution to its *Daphnia* in the manner demonstrated by your teacher. Have the other pair in your group add a drop of weak cola solution to its *Daphnia*. Let the slides sit for about 2 minutes.

10 Have one partner in each pair record the *Daphnia*'s heartbeats in a box on Student Sheet 9.2a, while the other partner keeps time for 10 seconds.

11 Switch roles and repeat Step 10. Calculate the average number of heartbeats per minute for the two trials and record it on the data table.

12 Exchange information with your group members.

13 Follow your teacher's directions for returning the *Daphnia* to its container. With your group, update your organism photo card for *Daphnia*.

REFLECTING
ON WHAT
YOU'VE DONE

① Based on what you have learned in this lesson, answer the following questions in your science notebook:

A. What effect does cola have on the heart rate of *Daphnia*? Explain.

B. What effect does alcohol have on the heart rate of *Daphnia*? Explain.

C. How would you expect a person's heart rate to change if he or she drank a large quantity of cola or alcohol?

② Refer to the reading selection "The Transparent Water Flea" on pages 132–133 to respond to the following in your science notebook:

A. List three ways in which *Daphnia* are similar to other crustaceans.

B. Explain why *Daphnia* are referred to as "water fleas."

C. Explain one function of the bristles on a *Daphnia*'s legs.

③ Revise as necessary the responses you made during "Getting Started." Discuss your changes with the class.

The Transparent Water Flea

Like the WOWBug, *Daphnia* belong to the phylum of jointed-limbed organisms called Arthropods. Like shrimp, crabs, and lobsters, *Daphnia* are members of a class of Arthropods called crustaceans. Crustaceans are characterized by an external skeleton, gills for gas exchange, two pairs of antennae, and numerous jointed appendages. Different species of *Daphnia* range in length from around 0.2 millimeters (0.008 inch) to more than 5 millimeters (0.2 inch).

Daphnia are particularly interesting animals to study because their exoskeletons, or carapaces, are transparent. This makes it easy to observe and identify their internal organs with a hand lens or a microscope. Their hearts beat very rapidly, pumping blood throughout their bodies. Their intestine, tubular in shape, extends from mouth to anus. Female *Daphnia* have a large brood chamber just below their heart. The brood chamber holds the female's eggs.

Daphnia have been called "water fleas" because they move with a jerky motion that resembles the way a flea jumps. They do this by quickly flipping their antennae downward. They control their depth in the water by adjusting the movement of their antennae like a parachute.

Daphnia can survive in almost any freshwater environment—lakes, ponds, streams, swamps, and marshes. They feed on microscopic organisms such as bacteria, algae, and protozoa. *Daphnia* propel food toward their mouths using water currents they generate with their leg movements. They filter out food particles with the bristles on their legs, then pass the food from the bristles to their mouths.

In their lifetime, female *Daphnia* produce up to 400 eggs. Reproduction in *Daphnia* follows an unusual process—the eggs develop in the female's brood chamber without being fertilized. Offspring are fully developed when they hatch.

Because *Daphnia* reproduce so rapidly, they are an important source of food for many other organisms, especially fish. They provide an important link in the food chain between the microscopic organisms upon which they prey and the larger organisms that prey upon them. ■

DISCUSSION QUESTIONS

1. Why are *Daphnia* easy to study and what has it allowed us to learn about them?

2. Why don't we study only organisms that are easy to study?

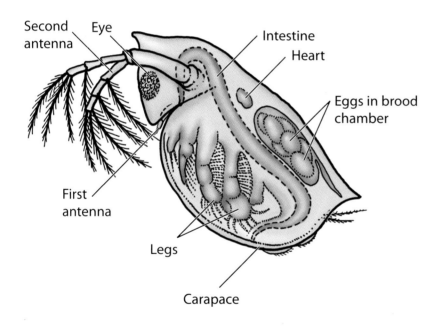

Second antenna
Eye
Intestine
Heart
Eggs in brood chamber
First antenna
Legs
Carapace

EXPLORING THE *HYDRA*

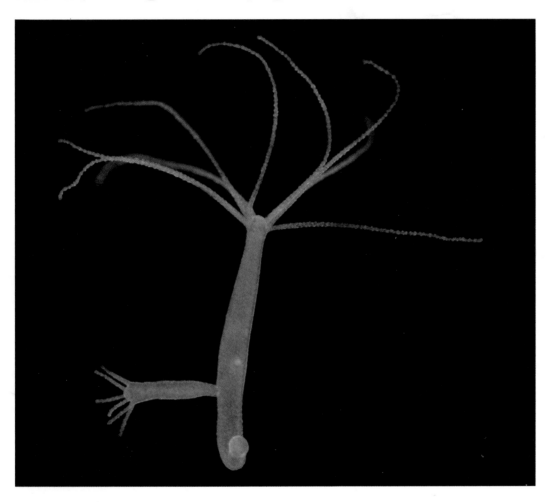

▶ **A GREAT VARIETY OF ORGANISMS, INCLUDING** *HYDRA*, **MAY BE FOUND IN A SMALL QUANTITY OF POND WATER.**

PHOTO: Peter J. Bryant, University of California, Irvine

INTRODUCTION

When you hear the word "*Hydra*," you may think of the nine-headed mythological beast with poisonous breath slain by Hercules. When you observe a multicellular *Hydra* under the microscope, you'll better understand how it got its name. The *Hydra* is a relative of the more familiar, and often feared, jellyfish. In this lesson, you will observe, sketch, and measure a *Hydra*. You also will observe the way it obtains its food, reacts to touch, and reproduces asexually.

OBJECTIVES FOR THIS LESSON

▶ Observe, sketch, and measure a *Hydra* and label its prominent features.

▶ Observe a *Hydra*'s methods of obtaining food and reacting to touch.

▶ Observe a *Hydra*'s method of asexual reproduction.

▶ Update the organism photo card for *Hydra*.

▶ MATERIALS FOR LESSON 10

For you

Your copy of Student Sheet 2.3a: Guidelines for Scientific Drawings

1 copy of Student Sheet 10.1: Template for *Hydra* Drawing

1 copy of Student Sheet 10.3: Template for *Hydra* Budding Drawing

For your group

1 set of organism photo cards

2 *Hydra*

1 prepared slide of *Hydra* budding

1 blackworm fragment

1 *Daphnia*

2 compound light microscopes

2 plastic depression slides

2 dissecting needles

2 metric rulers, 30 cm (12 in.)

2 transparent rulers

1 black marker

GETTING STARTED

1 One student from each pair must take the depression slide to your teacher to obtain a *Hydra*. Do not place a coverslip on your slide.

2 At your seat, take turns observing the *Hydra* under 40x. Then respond to the following in your science notebook: 📖

 A. Describe the *Hydra* in one or two sentences.

 B. List any organisms you think are similar to the *Hydra*.

 C. How do you think the *Hydra* obtains its food?

3 Discuss your observations and responses with the class.

4 Save your slide for use in Inquiry 10.1.

Based on this illustration, how do you think the *Hydra* got its name? Which of these dictionary definitions applies to the *Hydra* in this lesson?

1. a nine-headed serpent or monster of Greek mythology slain by Hercules, each head of which when cut off is replaced by two others

2. a multifarious evil not to be overcome by a single effort

3. a southern constellation of great length that lies south of Cancer, Sextans, Corvus, and Virgo and is represented on old maps by a serpent

4. any of numerous small tubular freshwater hydrozoans (as of the genus *Hydra*)

OBSERVING AND SKETCHING A *HYDRA*

PROCEDURE

1 Focus on a *Hydra* under 40x. Slowly move the slide so you can see the entire organism.

2 In the upper circle on Student Sheet 10.1: Template for *Hydra* Drawing, draw the entire *Hydra* in detail. Refer to "*Hydra*: Up Close and Personal" on pages 140–143 for labeling information.

3 Place the transparent ruler underneath the slide. Position the ruler so that you can measure the *Hydra*'s length. Record the length in the appropriate place on your drawing.

4 While looking through the microscope, gently touch a tentacle with the tip of a dissecting needle. Discuss with your partner the reaction of the *Hydra* to your touch. Discuss the *Hydra*'s reaction speed compared with how fast you would react if you were touched with the point of a pin.

5 Move on to Inquiry 10.2, using the same *Hydra* and slide.

INQUIRY 10.2

FEEDING THE *HYDRA*

PROCEDURE

1 One student from each pair should bring the *Hydra* slide to the materials station to obtain either a *Daphnia* or a blackworm fragment from your teacher.

2 Focus on the *Hydra* under 40x. If you have a *Daphnia*, keep watching until it brushes against the tentacles of the *Hydra*. If you have a blackworm, use the tip of your dissecting needle to nudge it toward the *Hydra*. Refer to the reading selections at the end of this lesson if necessary to respond to the following questions on Student Sheet 10.1:

A. How does the *Hydra* behave when the *Daphnia* or blackworm touches its tentacles?

B. How do you think the *Hydra* is able to trap organisms that are so much larger than it is?

C. How does the *Hydra* take the organism into its body?

3 Follow your teacher's directions for returning the organisms.

SKETCHING A BUDDING *HYDRA*

PROCEDURE

1 Place the prepared slide of *Hydra* on your microscope stage and focus on it under 40x.

2 Prepare a quick sketch of the entire organism in the circle on Student Sheet 10.3: Template for *Hydra* Budding Drawing. Label a tentacle, the adult *Hydra*, and a bud.

3 Return the materials to the designated area.

4 Work with your group to update the organism photo card with new information learned in this lesson.

5 Respond to the following questions on Student Sheet 10.3:

A. What did you notice about this *Hydra* that is different from the live one you observed?

B. How do you explain this difference in structure?

C. How might budding be an advantageous method of reproduction?

REFLECTING
ON WHAT
YOU'VE DONE

1 Based on what you have learned in the lesson, respond to the following questions in your science notebook. You should be prepared to discuss them with the class.

A. Why do you think the *Hydra* is considered one of the simpler multicellular organisms?

B. Which well-defined body system(s) did you observe in the blackworm that you did not notice in the *Hydra*?

C. Which reproductive process do *Hydra* and blackworms have in common?

D. Based on the information in "Jellyfish Get a Bad Rap" on pages 144–147 list two ways in which *Hydra* and jellyfish are alike and two ways in which they are different.

HYDRA: UP CLOSE AND PERSONAL

Hydra, one of the simplest multicellular organisms, is a member of the phylum Cnidaria. *Hydra* are freshwater animals. Their bodies are thin, hollow cylinders with five to seven tentacles extending from the mouth. *Hydra* come in many different colors, including tan, gray, green, and brown. Adult *Hydra* are typically 6 to 13 millimeters (0.2 to 0.5 inches) long and are capable of stretching out or contracting.

The most common method of reproduction for *Hydra* is budding, a form of asexual reproduction. A small growth, or bud, forms on the adult *Hydra* body through a series of cell divisions. The bud soon develops tentacles and breaks away. The new *Hydra* that breaks off can then live independently.

Hydra also can reproduce by regeneration, another form of asexual reproduction. They can grow new bodies from small pieces that have been detached. When fall arrives and cooler conditions prevail, *Hydra* usually develop sex organs and reproduce sexually. This is because the eggs that are produced are able to survive the colder conditions, while the parents cannot.

Hydra have no circulatory systems, no heart or blood vessels. Because their bodies are only two cell-layers thick, oxygen simply drifts, or diffuses, into areas of the cells that are low in oxygen. This occurs as *Hydra* move through the water and their cells interact with the environment, exchanging oxygen for carbon dioxide through the cell membranes.

Hydra have no skin as we would think of it, but the outer layer of cells protects the organism. Nor does it have what we would think of as a stomach, but the inner layer of cells secretes enzymes that digest food, which comes in through a primitive mouth opening.

Hydra do not have a central nervous system. Instead, they have a "nerve net" through which impulses that control muscle contractions are carried. These contractions allow *Hydra* to expand, contract, and move about.

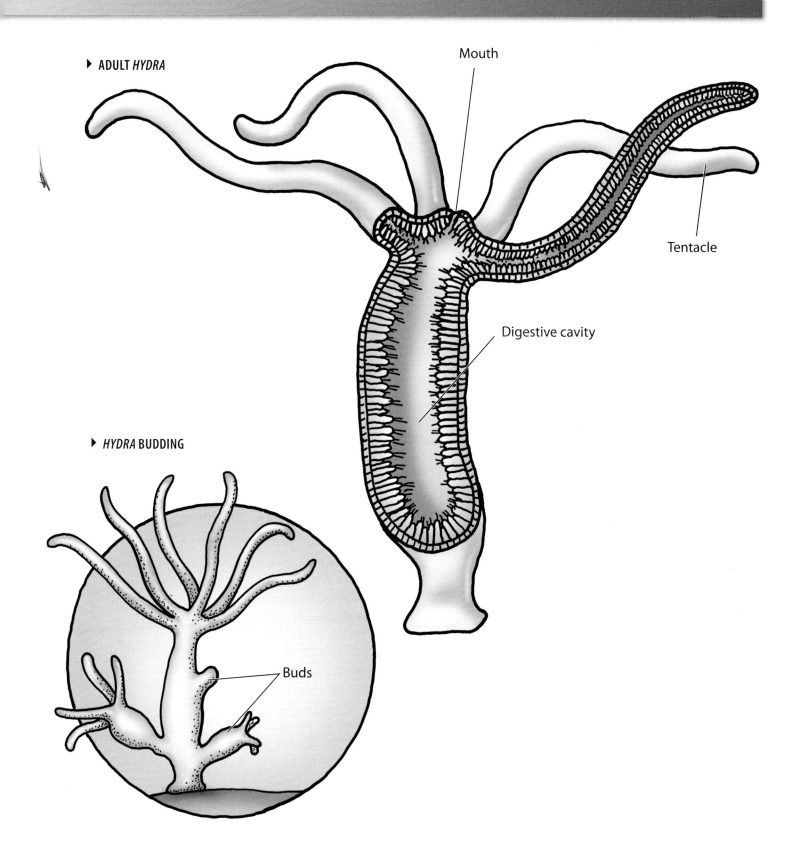

▶ ADULT *HYDRA*

Mouth

Tentacle

Digestive cavity

▶ *HYDRA* BUDDING

Buds

READING SELECTION

EXTENDING YOUR KNOWLEDGE

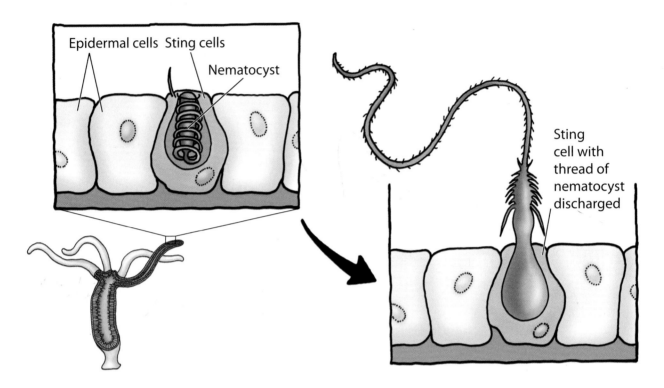

Epidermal cells Sting cells

Nematocyst

Sting cell with thread of nematocyst discharged

▶ NOTICE THE COILED THREAD INSIDE THE STING CELL ON THE LEFT. IT HAS BEEN DISCHARGED FROM THE STING CELL ON THE RIGHT.

PARALYZING POISONS AND TIGHT EMBRACES

Hydra feed on small organisms, such as *Daphnia.* Specialized cells that make up part of the outer layer of *Hydra* contain stinging structures called nematocysts. When an organism such as *Daphnia* brushes against *Hydra*'s tentacles, the nematocysts shoot out threads. These threads pierce the prey and release a paralyzing poison. Other nematocysts release threads that surround the prey and hold it tight. The tentacles then move the prey toward *Hydra*'s mouth and force it into its digestive cavity.

Hydra secrete digestive enzymes into this cavity from cells in the inner layer. These enzymes break down the prey into usable nutrients. Whip-like flagella, which line the digestive cavity, wave about to stir the digestive fluid. Undigested pieces are released through *Hydra*'s mouth. Digested nutrients move from the cavity into the cells of *Hydra* by diffusion.

Hydra spend much of their time attached to a surface by their bases. However, when it's time to move, *Hydra* move in style. They simply float or travel from one location to another by doing somersaults! ■

▶ THE *HYDRA* IS SOMEWHAT OF A GYMNAST, OFTEN PERFORMING SOMERSAULTS TO MOVE FROM ONE PLACE TO ANOTHER.

▶ THIS *HYDRA* HAS CAPTURED AND PARALYZED A MINNOW. IS IT BITING OFF MORE THAN IT CAN CHEW?

PHOTO: Courtesy of Carolina Biological Supply Company

DISCUSSION QUESTIONS

1. A *Hydra* is likely to be found rooted to a particular spot, looking more like a statue or plant than an animal. What characteristics prove that it is in fact alive, and an animal?

2. How many different ways can *Hydra* reproduce and what is the advantage of having several ways?

Jellyfish Get a Bad Rap

Consider jellyfish. They may look like jelly, but they're not really fish at all. In fact, they are close relatives of *Hydra*. Their reputation for stinging has made them very unpopular. And it's really not fair. They're poor swimmers and often bump into things—including people. When jellyfish sense movement nearby, barbed, threadlike stingers automatically shoot out from their tentacles. It's a good strategy for stunning and killing small fish and other prey—but not the best for making friends.

Luckily, most people who get stung by jellyfish don't meet the fate of those prey. They develop a red, itchy rash that goes away in a few hours. Putting ice on a sting often makes it feel better. That is, unless you get stung by a Portuguese man-of-war, a type of jellyfish found in warm seas. Then, look out for painful welts and a fever—and maybe a trip to the hospital.

The Portuguese man-of-war, while impressive, is perhaps not the most dangerous jellyfish. In the warm waters of the Great Barrier Reef of Australia, a relatively small species, the box jellyfish, terrorizes swimmers. Its poison is extremely potent and life-threatening. At some popular beaches, jellyfish nets are installed around the perimeter of swimming areas to protect people from encounters with these organisms.

As for jellyfish you may find lifeless on the beach—beware! Stingers can shoot out for a few hours even after a jellyfish has died.

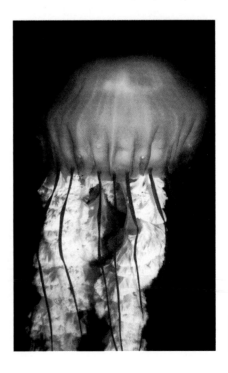

▶ **BOX JELLYFISH**
PHOTO: U.S. Marine Corps

▶ JELLYFISH ARE LEFT BEHIND
WHEN THE TIDE GOES OUT
IN MELBOURNE, AUSTRALIA.
THEY MAY STILL STING FOR A
SHORT TIME AFTER THEY DIE.

PHOTO: Andrew Braithwaite/
creativecommons.org

READING SELECTION

EXTENDING YOUR KNOWLEDGE

MOSTLY WATER

Jellyfish are invertebrates, meaning they have no backbone. But that's not the half of it. Jellyfish also have no brain, heart, blood, bones, eyes, ears, or gills. In fact, they're 95 percent water. Most are bell-shaped, and through their hollow transparent bodies you can see—if you care to—what they had for lunch.

Jellyfish can alternately contract and relax their bell-shaped bodies to push themselves along in the water. But much of the time they simply drift along on ocean currents, with their tentacles—which can range in length from only 1 centimeter (0.4 inches) to as much as 30 meters (98.4 feet)—trailing below them.

Considering how ancient they are, jellyfish get very little respect. These ocean dwellers have been floating around since long before the dinosaurs! And it's not as if they have it easy. Many kinds of fish, as well as sea turtles and marine birds, seek them out for food.

Jellyfish can be found floating freely in the ocean, but they can also be found with their cousins, the corals. Corals are cnidarians that form a hard outer skeleton from calcium carbonate, and in colonies they form enormous, rocky reefs that are home to an astonishing array of diversity and color in the oceans. Jellyfish living among the corals are prey to reef-dwelling turtles and other animals, but they find room to reproduce there; jellyfish are male and female, reproducing sexually, and forming tiny polyps that root on the coral bed and grow into small jellyfish. In the reef, they can feed off small fish and other creatures, living and dead, that drift by and bump into them.

These reef habitats are threatened by climate change. As the oceans grow warmer and more acidic, many coral reefs bleach, actually turning white and losing much of the life that swarms around them. That's because the coral reef ecosystems depend on tiny, photosynthesizing algae that live among the corals. When the water warms or its chemistry changes, the coral can expel their algae. Then the reefs appear white, and the first links in the food chain are lost. Organisms that eat the algae find no way to live there, and larger organisms that eat those creatures must also leave or starve.

NUMBERS OUT OF WHACK

Even so, the mobile jellyfish thrive—in fact, they're doing better than ever, finding homes among human-built dock structures that are reef-like enough for them. They're also tolerant of higher sea temperatures. In some areas, they have become so numerous and eat so much that fish, shrimp, crab, and other seafood are in short supply. This threatens the livelihood of people around the Gulf of Mexico and elsewhere who make their living by fishing. Some are even growing to monstrous sizes; recently a Japanese fishing boat accidentally picked up a net full of 180-kilogram (400-pound) jellyfish, which sunk the boat as the fishermen tried to lift it.

But again, you can't blame jellyfish. Their numbers increase when oxygen levels in the water are low, something that happens when a lot of fertilizer and waste products get dumped into the water. In addition, in some areas overfishing has left jellyfish with few predators. ∎

▶ THE CORAL REEFS THAT MANY JELLYFISH RELY ON
ARE THREATENED BY HUMAN ACTIVITIES.

PHOTO: Mr. Mohammed Al Momany, Aqaba, Jordan/NOAA's Coral
Kingdom Collection

DISCUSSION QUESTIONS

1. What are some characteristics of jellyfish that have allowed them to persist since before the time of the dinosaurs?

2. Jellyfish die when they get stranded on beaches. What additional characteristics would a jellyfish need to be able to live on land?

INVESTIGATING FUNGI I—THE MOLDS

▶ **FUNGI COME IN MANY SHAPES AND SIZES. HERE IS ONE EXAMPLE.**

PHOTO: Tim Parkinson (flickr. com/timparkinson/)/Creative Commons Attribution License (http://creativecommons.org/ licenses/by/2.0)

INTRODUCTION

What do you think of when you hear the term *fungus*? Does it conjure up a positive image in your mind? Or is it something you normally associate with things that aren't so pleasant? Actually, fungi (the plural of fungus) are very important to us in many ways. In this lesson and the next, you will explore the nature of fungi and the important role they play in our lives.

OBJECTIVES FOR THIS LESSON

Observe a photo of "mystery prints" and agree on how they were formed.

Decide on conditions favorable for the formation of mold.

Compare the rate of mold formation on two types of bread.

Observe and document the progress of a fungal garden.

Update the bread mold organism photo card.

▶ MATERIALS FOR LESSON 11

For your group

1	copy of Student Sheet 11.1: Comparing Mold Formation on Two Types of Bread
1	set of organism photo cards
2	small resealable plastic bags
1/4	slice of brand-name bread
1/4	slice of freshly baked or homemade bread
1/2	paper towel
1	pair of scissors
2	hand lenses
1	plastic pipette
1	black marker
	Transparent tape
	Tap water

GETTING STARTED

1 With your group, observe Figure 11.1. It shows a "mystery print" left on a piece of paper by a living organism that is probably familiar to most of you. Agree on the type of organism that left the print and how the print was formed. List this information in your science notebook.

2 Share your ideas with the class while your teacher records your responses.

3 Agree on and record three conditions you think would be favorable for the growth of mold.

4 Share your ideas with the class. As a class, narrow your ideas to the three most favorable conditions. Next, identify places in your classroom where these conditions are most likely to exist.

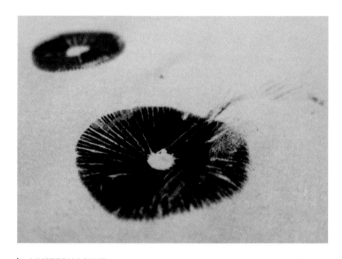

▶ **MYSTERY PRINT**
FIGURE **11.1**

PHOTO: gr33n3gg/creativecommons.org

SAFETY TIP

When culturing live organisms, such as mold, always keep them in a sealed container.

Return all sealed plastic bags containing mold to your teacher for disposal.

Wash your hands after handling moldy food.

COMPARING MOLD FORMATION ON TWO TYPES OF BREAD

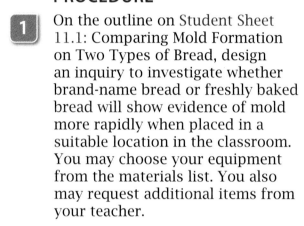

PROCEDURE

1 On the outline on Student Sheet 11.1: Comparing Mold Formation on Two Types of Bread, design an inquiry to investigate whether brand-name bread or freshly baked bread will show evidence of mold more rapidly when placed in a suitable location in the classroom. You may choose your equipment from the materials list. You also may request additional items from your teacher.

2 While designing your inquiry, discuss and agree on what the inquiry should include in order to yield valid results. Consider the following questions:

- How will you ensure that both types of bread are kept under the same conditions?

- What kind of results are you looking for, and how will you measure them?

- How often will you check your bread and record your observations?

- How and where will you record your observations?

- What items of equipment will you need?

- How will you distinguish your bread from that of other groups in your class?

- How will you communicate your results to the rest of the class?

- What will you do to make sure your inquiry meets the safety requirements for handling moldy substances?

Inquiry 11.1 continued

INQUIRY 11.2

3 Include in your inquiry design a list of at least four things that you will keep constant in your inquiry.

4 Set up your materials for the inquiry. Return any unused equipment to your teacher.

5 Once you've completed your inquiry, record your results on Student Sheet 11.1. Then, follow your teacher's directions for cleaning up.

6 With your group, update your organism photo card for bread mold.

▶ **MOLD HELPS DECOMPOSE ORGANIC MATTER, SUCH AS THIS DEAD CRICKET.**

PHOTO: Courtesy of Carolina Biological Supply Company

CREATING AND OBSERVING A FUNGAL GARDEN

PROCEDURE

1 Your teacher has placed in your classroom a plastic container filled with food brought in by volunteers. Observe the container's contents every two days for evidence of mold.

2 Follow your teacher's directions for recording your observations of the container's contents in your science notebook until he or she tells you it is no longer necessary. Include such things as: 🖎

- The date each of the foods first showed signs of molding

- Whether there are different kinds of mold in the container

- Whether there are signs of organisms other than mold in the container

▶ **WATCH OUT FOR AMANITA! IT'S ONE OF THE WORLD'S MOST BEAUTIFUL FUNGI, BUT IT CAN BE DEADLY IF EATEN.**

PHOTO: Courtesy of Carolina Biological Supply Company

▶ **IT'S EASY TO SEE HOW THIS BIRD'S NEST FUNGUS GOT ITS NAME. THE "EGGS" ARE ACTUALLY MASSES OF SPORES THAT SPLASH OUT OF THE "NEST" WHEN BOMBARDED BY A RAINDROP.**

PHOTO: Courtesy of Carolina Biological Supply Company

REFLECTING ON WHAT YOU'VE DONE

1. Answer the following questions in your science notebook. For assistance, refer to the reading selection "There's a Fungus Among Us" on pages 154–157.

A. Which of the two bread samples first showed signs of molding? How do you account for this?

B. Why do you think fungal inhibitors are added to bread if the bread will mold anyway?

C. Explain why bread mold may be present on a piece of bread well before you might notice it.

D. Were the room conditions you chose the best for mold formation? Did you prove this? How?

E. Was there more than one species of mold in your fungal garden? How could you tell?

F. Most fungi are "decomposers." What does this term mean, and why are decomposers so important to us? How does this relate to what happened in the fungal garden?

G. Why do different kinds of food items decompose at different rates?

READING SELECTION
EXTENDING YOUR KNOWLEDGE

THERE'S A FUNGUS AMONG US

Fungi are strange organisms. Unlike animals, they can't capture their own food. And unlike plants, they can't make food from sunlight. Most can't even move around.

But fungi are hardy organisms. They are found almost everywhere. Some are deadly if eaten, but they also can play a positive role in cleaning up our environment.

WHAT BELONGS IN THE FUNGI KINGDOM?

The Fungi kingdom is huge. Some scientists estimate there are as many as 1.5 million species of fungi. These species may be unicellular or multicellular, and all break down organic material to get the nutrients they need. About 100,000 have been identified so far. Fungi range in size from microscopic to very large. One common species, black bread mold, grows on bread, fruit, and other foods. It's fuzzy and has tiny black dots when viewed without magnification. The underground parts of one particular species of mold are so extensive that it is considered to be one of the largest organisms on Earth. In addition to molds, fungi include mushrooms, puffballs, yeasts, lichen, and most mildews.

ANATOMY OF A FUNGUS

The body of a typical fungus is made up of many tiny tubes, called hyphae. The hyphae are tangled into a thick mass, or mycelium. These structures help the fungus absorb and digest its food. Because they can't move about, fungi usually live where they eat—right on their food source. As the mycelium grows, it covers larger portions of its food source. The fungus releases strong juices that break down plant and animal matter. The cells of the hyphae then absorb this matter.

The most visible part of a fungus is its reproductive structure. This is especially true for club fungi, which include the mushroom. Much of a mushroom's mycelium grows below ground, but its reproductive structure—the familiar umbrella-shaped cap—sprouts from the ground, supported by a thick stalk.

For fungi, tiny structures called spores play a role similar to that of seeds in plants. Most fungi produce large numbers of spores. The spores are located inside the mushroom's cap. When released, the spores float away. Most of them die because they land in places that are not the right temperature or do not have enough food or moisture to support growth. When they land in a suitable spot, however, the spores germinate and grow.

▶ THERE IS AN AMAZING VARIETY OF SPECIES IN THE FUNGI KINGDOM. THE STRUCTURES WE ARE ABLE TO SEE ARE USUALLY THE REPRODUCTIVE ONES.

PHOTO (top left): polandeze/creativecommons.org
PHOTO (right): Anssi Koskinen/creativecommons.org
PHOTO (bottom left): Courtesy of Carolina Biological Supply Company

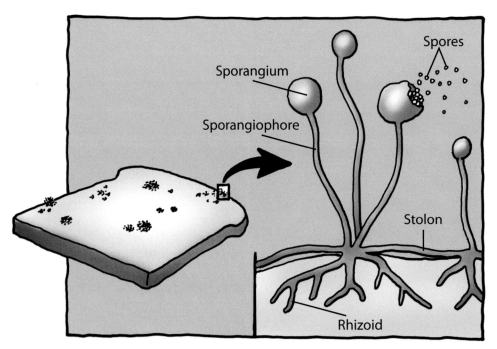

▶ **DEVELOPMENT AND STRUCTURE OF BREAD MOLD**

Ever notice how mold seems to come from nowhere? That's because tiny mold spores, like those of the mushroom, travel by floating in the air. When a bread mold spore lands on a piece of moist bread, it grows specialized hyphae called rhizoids, which travel below the surface of the bread to help anchor the mold, like roots anchor a plant. Other hyphae spread along the surface of the bread, forming a mycelium. Other specialized hyphae, called stolons, develop from the mycelium. Long stalks soon grow upward from the stolons. Round reproductive structures called sporangia form at the top. New spores form inside the sporangia. Sporangia give molds their characteristic colors. The black color of bread mold, for example, becomes apparent only as the sporangia develop. That's when you notice you've got moldy bread.

FORMING RELATIONSHIPS

Some fungi are symbiotic, which means that they exist in a long-term relationship with another organism, usually to their mutual benefit. For example, Central American leaf-cutter ants actually cultivate fungi and plant their spores on leaves and flowers that they have chewed into small bits. Later, the ants eat the fungi that develop from those spores. Lichens consist of a fungus and an alga living together, each getting some of its nutrients from the other.

Some fungi are not so friendly. They are parasitic, which means that they live on the body of a plant or animal and use it as their food source. These fungi eventually can cause great harm to the plant or animal on which they feed.

▶ **THE BRITISH SOLDIER LICHEN GETS ITS NAME FROM ITS BRIGHT RED CAP.**

PHOTO: Sandy Richard

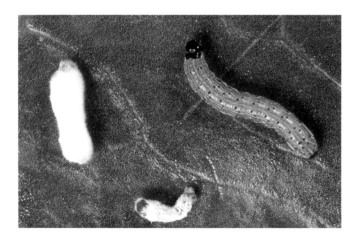

▶ **THE TWO CATERPILLARS ON THE LEFT WERE KILLED BY A FUNGUS.**

PHOTO: Keith Weller, Agricultural Research Service/U.S. Department of Agriculture

These parasitic relationships are very specific. One species of fungus that feeds on elm trees, for example, causes Dutch elm disease. It has killed millions of elm trees, but it does not attack any other type of tree.

WHAT GOOD ARE FUNGI?

Fungi have many useful roles. For example, they are excellent recyclers. Fungi break down plant and animal remains and wastes and release their nutrients back to the Earth. You might even think of them as our planet's master decomposers. Because of these capabilities, fungi have been called into action on environmental problems. Scientists use them to eliminate certain insect pests and to decompose—and render harmless—some of the pesticides and other chemicals that pollute our water and our soil.

Some types of mold are used to make antibiotics, which are medicines that fight bacterial infections. Other molds ripen cheeses, such as Brie and Roquefort. Yeast, another type of fungus, helps make wine from grape juice and is added to dough to make bread.

While some of these organisms won't win any beauty contests, they're mighty important to our lives—and to our world. ■

 DISCUSSION QUESTIONS

1. Why not include fungi in the Plant or Animal kingdoms?

2. Describe and explain what you are eating if you eat a piece of moldy bread.

INVESTIGATING FUNGI II—YEAST

▶ WHAT DO YOU THINK
THIS PHOTO HAS TO
DO WITH THE TOPIC
OF THIS LESSON?

PHOTO: César Astudillo/
creativecommons.org

INTRODUCTION

In this lesson, you will focus on another member of the Fungi kingdom—yeast. You may have heard of yeast being added to bread or cake dough to make it rise. But did you know that yeast cells are living organisms? In this lesson, you will observe a yeast culture and watch as evidence of an important life process bubbles up before your eyes. You will design an inquiry to investigate substances that affect yeast cell activity. You also will read about the important role that yeast plays in our daily lives.

OBJECTIVES FOR THIS LESSON

Observe evidence of yeast activity.

Design and conduct an inquiry to investigate substances that will or will not promote yeast activity.

Explain how different kinds of yeasts benefit or harm humans.

Update the yeast organism photo card.

▶ MATERIALS FOR LESSON 12

For you

1   copy of Student Sheet 12.1: Investigating the Effect of Two Substances on Yeast Activity

For your group

1   set of organism photo cards
2   metric rulers, 30 cm (12 in.)
3   test tubes, 25 mm x 150 mm
1   test tube rack
1   10-mL graduated cylinder
1   250-mL beaker
1   black marker
3   stirrers

GETTING STARTED

1 With the class, read "Introducing Yeast." Then observe and participate as your teacher provides more information about yeast.

2 In your science notebook, divide the list of substances your teacher shows you into two groups. In the first group, include the substances that you predict will promote yeast activity. In the second group, include those substances that you predict will not promote yeast activity. ☑

3 Share your lists with the class.

INQUIRY 12.1

EXPLORING YEAST CELL ACTIVITY

PROCEDURE FOR PERIOD 1

1 Choose one substance from each of the groups you made during "Getting Started."

2 Work with your group to devise and outline an inquiry on Student Sheet 12.1: Investigating the Effect of Two Substances on Yeast Activity to investigate whether the two substances you selected were placed in the correct group. For Step B (What I think will happen and why), write one sentence about each of the two substances, telling why you placed each in its group and what you think will happen when they are mixed with yeast cells. Be sure to include a valid control and an explanation of when and how you will measure your results. Your outline should include items A–F in the list that follows. Your final inquiry should include items A–H, which can also be found on your student sheet.

A. Question I will try to answer

B. What I think will happen and why

C. Materials I will use

D. At least four things I will keep the same (you may list more)

E. Procedure I will follow

F. Data table

G. Graph of my findings

H. What I found out

3 After you complete your inquiry design, read "Yeast: Rising to the Occasion" on pages 163–165.

INTRODUCING YEAST

There are many species of yeasts. They are divided among three different phyla of the Fungi kingdom. Most yeast species, however, belong to the phylum Ascomycetes. In addition to yeasts, this phylum includes truffles, morels, and mildew. Most fungi are multicellular and relatively large. Yeast cells are unusual because they are unicellular and microscopic. Scientists believe that yeast once had the typical fungi's ability to form hyphae—the tubes that root fungi to the surface of an object—but gradually lost that ability.

Dry granules of yeast contain tiny spore sacs. In a moist, warm environment in which a food source is available, the spores become active; during this period, they grow into new yeast organisms and begin to reproduce. Although yeast cells can reproduce sexually, they usually reproduce asexually through a form of cell division called budding. In this process, a new cell forms by cell division and produces a small outgrowth on an older cell. Eventually, the smaller cell breaks off and becomes self-sufficient.

When yeast cells become active and feed, they undergo a process called fermentation. During this process, sugar is broken down and carbon dioxide and alcohol are formed. In this lesson, you will see and measure evidence of yeast activity as yeast grains are added to different substances. ■

▶ NINETEENTH-CENTURY MICROBIOLOGIST LOUIS PASTEUR (1822–1895) FIRST CULTIVATED YEAST CELLS AND USED THEM FOR SCIENTIFIC PURPOSES.

PHOTO: Library of Congress, Prints & Photographs Division, LC-USZ62-3499

▶ YEAST GRANULES

PHOTO: ©2010 Carolina Biological Supply Company

Inquiry 12.1 continued

PROCEDURE FOR PERIOD 2

1. During this period, you will conduct the inquiry that you designed. Use the black marker to label each of your three test tubes with the name of the substance you are testing.

2. At the appropriate time, complete and record any necessary measurements. Exchange information with other groups until you have data for all of the substances. If other groups tested the same substances as your group, average their data with yours and record only the average.

3. Follow your teacher's directions for cleaning up and returning your materials. Graph your findings and complete Step H on your student sheet.

4. Update your group's organism photo card for yeast.

REFLECTING
ON WHAT
YOU'VE DONE

1. Answer the following questions in your science notebook, then discuss your responses with the class:

   A. If the foam column for either of the mixtures you tested was not as high as the column produced in the sugar solution, would you classify that substance as a promoter or non-promoter of yeast activity? Explain.

   B. Were the predictions you made in "Getting Started" correct? Explain.

   C. Were your results consistent with those of other students who tested the same substances? If not, explain why you think they differed.

   D. Did any of your results surprise you? If so, which results and why?

2. Refer to "Yeast: Rising to the Occasion" to respond to the following in your science notebook:

   A. Explain one way in which yeast can be harmful to humans.

   B. Explain two ways in which we use yeast to our advantage.

Yeast: Rising to the Occasion

Yeast cells sure do get around. You'd be surprised at all the places you can find them. These tiny, one-celled organisms live all around us—in soil and saltwater, on plant leaves and flowers.

Yeast are fungi. Like other fungi, yeast cells are very good at recycling. They stay busy by decomposing, or breaking down, plant and animal matter. As they do this, they grow and reproduce, and in the process, carbon dioxide and alcohol are released. They're capable of reproducing quickly, within minutes, by budding. When cells bud, they actually grow tiny new cells near their edges, complete with a copy of the parent cell's DNA. The buds pinch off, taking part of the parent cell's membrane for their own, and then grow to full size.

THE YEAST WITHIN

Yeast cells not only live all around us, they also live upon us and within us! The oily surfaces of our noses, ears, and scalps are favorite hangouts. So are our mouths and intestinal tracts. It may seem weird, but it's all perfectly normal.

Most of the time, the yeast populations on our bodies are present in numbers that cause no problems. Sometimes, such as when we take certain medications or change our diets, yeast colonies are able to multiply rapidly. This can lead to infections. When yeast cells overgrow in the lining of our mouths, for example, we call it thrush, an uncomfortable, contagious disease found most often among babies and children. Symptoms of thrush include fever and diarrhea, and small whitish bumps on the mouth, throat, and tongue.

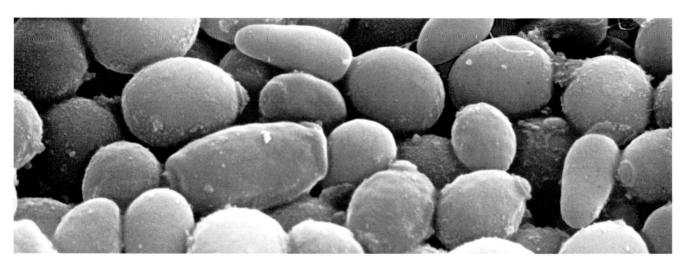

▶ **YEAST BUDDING**

PHOTO: by T.K. Maugel, Laboratory for Biological Ultrastructure, University of Maryland, College Park

READING SELECTION
EXTENDING YOUR KNOWLEDGE

▶ THE DOUGH USED TO MAKE THE BREAD IN THE UPPER PHOTO HAD NO YEAST ADDED. YEAST WAS ADDED TO THE DOUGH FOR THE BREAD IN THE LOWER PHOTO.

PHOTO (top): RonAlmog/creativecommons.org
PHOTO (right): Scott Bauer, Agricultural Research Service/U.S. Department of Agriculture

THE BREWER, THE BAKER

Most of us know yeast best from its role in breadmaking. When the conditions are right—warm and moist—yeast cells make the bread dough rise, or expand. They do so during two chemical processes called fermentation and respiration. Respiration is a process through which cells use oxygen to get energy by breaking down simple sugars; in fermentation, no oxygen is used for the breakdown.

Yeast that are kneaded into bread dough actually feed on the dough itself, reproduce, and excrete a gas, carbon dioxide, plus alcohol. This is *fermentation*. After a time, the yeast switch to eating the alcohol they've made, using oxygen to break it down and turn its energy into fuel for growth and reproduction. This is *respiration*. In turning the alcohol into fuel for themselves, they excrete more carbon dioxide.

That carbon dioxide accumulates in the bread, making it rise, and when the bread is baked the carbon dioxide expands. When you look closely at a piece of bread, you can see little holes where the carbon dioxide bubbles have been. When the bread is baked, the dough stays in this expanded shape. The heat from the oven also kills the yeast organisms. So the gas that makes your bread light and airy is essentially…yeast poop!

Ethanol is also an important part of beer and wines, and indeed the same yeast, *Saccharomyces cerevisiae*, is used in both breadmaking and beer brewing. The bubbles in certain alcoholic drinks, such as beer and champagne, come from the carbon dioxide that gets released when the grapes or malt ferment.

THE SCIENTIST'S MAINSTAY

As important as bread, beer, and wine are, yeast are important for considerably more than that. Yeast, which are larger than most bacteria and reproduce easily, are very useful in scientific research, and have been for hundreds of years. Research with yeast led to early understanding of how we burn sugar with oxygen for cellular fuel; yeast were used in developing pasteurization and the first vaccines; yeast taught us first about enzymes, the huge molecules that speed life's biochemical reactions. Yeast are still used today in research on genes and the life of eukaryotic cells.

There's no doubt about it—yeast cells are extremely busy, often useful, and occasionally irritating, organisms. ■

▶ THE POWERS OF YEAST TO MAKE ALCOHOL HAVE LONG BEEN KNOWN. HERE, CALIFORNIANS IN THE EARLY 1900S LOAD RAILROAD CARS WITH GRAPES BOUND FOR FERMENTATION AT WINERIES.

PHOTO: Library of Congress, Prints & Photographs Division, LC-USZ62-93513

DISCUSSION QUESTIONS

1. How is it correct to say that making bread means making alcohol? What happens to the alcohol?

2. We rely on yeasts for many foods. What is another group of organisms you learned about in this unit that we rely on heavily? How?

ORGANIZING LIVING THINGS—AN ASSESSMENT

▶ **THERE ARE SEVERAL DIFFERENT KINDS OF ORGANISMS IN THIS CORAL REEF. WHAT CHARACTERISTICS DO THEY SHARE? WHAT CHARACTERISTICS ARE DIFFERENT?**

PHOTO: David Burdick, NOAA's Coral Kingdom Collection

INTRODUCTION

In this lesson, you will review the information on the organism photo cards for the living things you have studied during this unit. You will use this information to compare those organisms using Venn diagrams. Your teacher will review the use of Venn diagrams with you during "Getting Started." You will practice constructing Venn diagrams that compare and contrast two organisms from two different kingdoms. Then you will create a Venn diagram for three different photo card organisms from three different kingdoms. Using your Venn diagrams, you will prepare written lists of the characteristics of each kingdom.

OBJECTIVES FOR THIS LESSON

▸ Develop Venn diagrams for pairs of various organisms selected from the organism photo cards.

▸ Create a Venn diagram for three photo card organisms from three different kingdoms.

▸ Develop a written list of characteristics of each of the kingdoms.

▸ **MATERIALS FOR LESSON 13**

For you

2   copies of Student Sheet 13.1: Venn Diagram for Two Organisms

1   copy of Student Sheet 13.2: Venn Diagram for Three Organisms

For your group

3   pieces of yarn, each approximately 1.8 m (72 in.)

1   set of organism photo cards

1   package of file cards

1   black marker

GETTING STARTED

1. Listen and participate as your teacher guides you through an activity using Venn diagrams. Your teacher will select two organisms from two different kingdoms for use in this activity.

2. Work with your partner to answer the following questions in your science notebook: ✐

   A. Which portion of the Venn diagram contains the features shared by both organisms?

   B. Which characteristics are unique to the first organism? (HINT: Look at the features that appear in the left-hand circle.)

   C. Which characteristics are possessed only by the second organism?

   D. List the features that are shared by both organisms.

3. Share your answers with the class.

▶ **HOW WOULD YOU COMPARE THESE ORGANISMS?**

PHOTO (right): John J. Mosesso/life.nbii.gov
PHOTO (left): J.E. Appleby/U.S. Fish and Wildlife Service

MAKING A VENN DIAGRAM

PROCEDURE

 1 Work with your partner to prepare a Venn diagram comparing two different organisms. Your teacher will select two organisms from among those represented by your organism photo cards. Follow these instructions to prepare your Venn diagram:

A. Obtain two pieces of yarn. Use one piece of yarn to form a large circle on your desktop. Use the second piece of yarn to form a second circle that slightly overlaps the first circle.

B. Write the name of the first organism on a file card. Place the file card just outside the lower edge of the first circle. On a second file card, write the name of the second organism and place it just outside the lower edge of the second circle.

C. With your partner, think of a characteristic shown by the first organism. Write this characteristic on a file card. (To help you think of features for each organism, you may refer to the notes you have made on your group's organism photo cards.) If the other organism also shares the characteristic, place the card in the area where the two circles overlap. If the characteristic is unique to that organism, however, place the card in the non-overlapping portion of that organism's circle.

D. Continue until you have selected at least five features or characteristics for each organism. As you are working, your teacher will review the features you have chosen. Check to make sure that some characteristics are unique to each organism, while both share other features.

E. Record the final Venn diagram on one of your copies of Student Sheet 13.1: Venn Diagram for Two Organisms.

F. Using the information in your Venn diagram, note the unique characteristics of the kingdom to which each organism belongs. Write these unique characteristics in your science notebook.

2 Repeat Step 1, this time using two different organisms as assigned by your teacher.

CREATING A VENN DIAGRAM FOR THREE ORGANISMS

PROCEDURE

1 Select three cards from the full set of organism photo cards. The organisms shown on the three cards must be members of three different kingdoms.

2 Using two of the organism photo cards, work with your group to follow Steps 1A and 1B of Inquiry 13.1.

3 Obtain a third piece of yarn. Use this yarn to add a third circle to your Venn diagram. This circle will represent the third organism photo card. Add a file card with the name of this organism to your diagram. Notice the centermost section of your Venn diagram. This section consists of the area in which all three circles overlap. Remember that you will use this area for characteristics shared by all three organisms.

4 Follow Steps 1C and 1D of Inquiry 13.1. When a characteristic is shared by one other organism, place the card in the area where the circles for the two organisms overlap. If a characteristic is shared by all three organisms, place the card in the central area where all three circles overlap. If you need help, ask your teacher to clarify anything you don't understand. Make sure you have placed at least one characteristic in every area of the Venn diagram. When you are finished, check with your teacher for approval.

5 Follow your teacher's directions for recording the Venn diagram on Student Sheet 13.2: Venn Diagram for Three Organisms, and for turning in your work.

6 Using the information in your Venn diagram, note the unique characteristics of the kingdom to which each organism belongs. Write these unique characteristics in your science notebook. 🖝

REFLECTING
ON WHAT
YOU'VE DONE

1 Answer the questions at the end of Student Sheet 13.2 and discuss your responses with the class.

2 Participate in a class discussion about the Venn diagrams you created. Be prepared to justify the characteristics you used and their placement in your Venn diagram.

3 Look back in your science notebook to your entries for Lesson 1. Look at the list of questions about organisms for which you wanted to find answers during this unit. On a fresh page, divide your questions into two groups: those that were answered and those that were not. Discuss both lists with the class.

4 Discuss with the class how you could reinforce or build on what you have learned in this unit.

5 Discuss with the class things that you learned in this unit that clarified some misconceptions you may have had about the organisms of different kingdoms.

TROPICAL RAINFORESTS: WHAT'S ALL THE HYPE?

If you walk into a tropical rainforest, you'll notice immediately that it's lush and very green. These unique forests have long gotten attention for the huge number of plants and other organisms that they harbor. However, in 1982, Smithsonian scientist Terry Erwin sparked a particular flurry of attention to the diversity of organisms in tropical rainforests through his work in Panama.

Shocking though it may seem, Erwin's method of alerting the world to the rainforest's biodiversity began with killing thousands of organisms. There was purpose to it, though. Terry decided to fumigate, or spray a fine mist of poison, over the top of a tropical rainforest, where the trees leaf out into a canopy. He pumped a fog of insecticides into the canopy of the rainforest. Insects killed by the insecticides fell down like rain to tarps that he had laid out below. He shocked people by finding almost a thousand different beetle species from just 19 trees. Judging from the huge variety of insects in such a small area, he estimated there to be 30 million species of organisms on earth. And humans are just one of those many species.

Scientists were excited by Terry's discovery and deepened their study of tropical rainforests.

Rather than bringing dead organisms to the ground, the scientists looked for ways of bringing themselves to the living creatures above. To get up into the trees, canopy researchers have used rock climbing equipment, construction cranes, hot-air balloons, ultra-light airplanes, and pulley systems like those used at ski lifts. The results have been astounding. For example, a single tree in Peru was found to harbor 43 species of ants. A single rainforest reserve in Peru contained more species of birds than the entire United States. The scientists found that one hectare of rainforest may contain nearly 500 tree species—compare that to the diversity in temperate forests like those found in the United States. Temperate forests are dominated by only 10 tree species. Although scientists have argued about the total number of species on Earth (current estimates range from 5 million to 30 million), it is agreed that rainforests are exceptionally diverse places.

What makes tropical rainforests so diverse? They have year-round high temperatures, lots of sunlight, and abundant rainfall. These are ideal conditions for many types of organisms, especially for the plants that form the base of the forest food web. The amount of

THE NETWORK OF SURFACE ROOTS ON THESE RAINFOREST TREES IN AUSTRALIA IS EVIDENCE OF THE THIN SOILS.

PHOTO: vaka0627/creativecommons.org

▶ THESE SCIENTISTS ARE SAMPLING LEAVES FROM THE TREE CANOPY. WOULD YOU LIKE TO HAVE THIS JOB?

PHOTO: Smithsonian Institution, Carl C. Hansen, #92-4630

photosynthesis going on in a tropical rainforest can be double what occurs in a temperate forest, thanks to the large leafy canopy that is exposed to lots of sunlight.

You might think these forests also have rich soils to support all the plants. On the contrary, they are unusual in having thin soils with few minerals. While the decay of dead organisms does produce minerals, these minerals are quickly washed through the soil by heavy rains. Rainforest plants must take up the minerals soon after they get released from decaying organisms. They also must remain

READING SELECTION

upright without much soil to support them. The spreading roots of a tropical rainforest tree serve both to access minerals and to prop up the tree.

Animals have specialized characteristics to live in tropical rainforests as well. If you see a column of ants marching with green umbrellas, those are the leaf-cutters. They chop up leaves and carry the pieces home to use for cultivating underground fungus gardens. Anteaters use their long, thin snouts to sniff out and eat the ants that are so abundant. Birds like toucans have stout, powerful beaks to crack open rainforest nuts. Tiny frogs live in puddles of water that collect in the leaves of rainforest plants. Emerald tree boas are camouflaged by the green foliage where they hunt for small prey. And by being nocturnal, many animals avoid predators that would threaten to eat them in the daylight hours.

This diversity of species makes tropical rainforests unique, but why should we try to protect the rainforests? One of the strongest arguments for saving rainforests has been that we need them for our own survival. It's estimated that the Amazon rainforests alone produce about 20% of Earth's oxygen gas. Recall, too, that plants take in carbon dioxide during photosynthesis. This removes carbon dioxide, a greenhouse gas, from the atmosphere. It's believed that the photosynthesis in tropical rainforests helps to slow global warming.

It also turns out that rainforest organisms are an important source of new chemicals that are used in medicine. About 25% of the medicines used in the U.S. have active ingredients that were first discovered in rainforest plants, and are now synthesized, or made, in laboratories. About 70% of the plants that have been identified by the U.S. Cancer Institute to be effective against cancer originated in rainforests. Consider Vincristine, which is used to treat childhood leukemia. It is extracted from a rainforest plant called the Madagascar periwinkle and has saved thousands of children's lives.

A tropical rainforest is like a giant library of organisms that may hold many useful materials for humans. Rainforest scientists and drug companies have collaborated to tap into the wealth of potential medicines that rainforest organisms contain. While the drug companies can provide money and expertise to test organisms for medicinal compounds, scientists have the equipment and techniques to gather the organisms. Indigenous people are another critical link in the process of discovering new medicines. They are familiar with the rainforests they live in and often have knowledge of the properties of organisms.

▶ PEOPLE WHO LIVE IN THE RAINFOREST OFTEN HAVE FIRST-HAND KNOWLEDGE OF THE PLANTS AND ANIMALS.

PHOTOS: Shea Hazarian/creativecommons.org

THESE SPECIES ARE SPECIALIZED FOR LIFE IN TROPICAL RAINFORESTS.

PHOTO (top): Alexander Torrenegra/creativecommons.org
PHOTO (bottom right): Alexander Torrenegra/creativecommons.org
PHOTO (right): Marcos Guerra/Smithsonian Tropical Research Institute
PHOTO (middle): Jessie Cohen, Smithsonian National Zoo
PHOTO (bottom left): psyberartist/creativecommons.org

READING SELECTION

EXTENDING YOUR KNOWLEDGE

▶ **THIS HILLTOP IN THAILAND SHOWS THE EFFECTS OF SLASH-AND-BURN PRACTICES.**

PHOTO: mattmangum/creativecommons.org

We are in a race against time: as quickly as new species are being discovered in the tropical rainforests of the earth, other species are going extinct. Rainforests, as recently as the 1950s, covered 15% of the land surface. They now cover only 6–7%. Scientists estimate that more than 100 species of organisms are going extinct every day due to the loss of rainforest. While rainforests may be home to as many as half of all the world's species of organisms, they could be gone entirely in 50 years.

Human needs are driving the destruction of rainforests. Growing numbers of people in the tropics have cut down rainforests to make a living as ranchers and farmers. The common method of clearing—called "slash-and-burn"—leaves little standing. Although people who live near rainforests put direct pressure on them, the overall impacts on rainforests are far from local. Large international corporations have replaced rainforests with banana, coffee, cattle, and other types of farms. Trees from rainforest lands are used by industries to provide products for sale, such as building materials, paper, and charcoal. Teak wood has become so popular for furniture manufacture and grows so fast that it has become a plantation tree in the tropics. In other words, natural rainforests are cut down and replaced with large-scale teak plantations to provide wood for exporting to other countries.

Are these tropical rainforest lands suitable for agriculture and ranching? No. Recall that rainforests have thin soils that do not store many nutrients. When they are burned and cleared, the nutrients, which were contained in the plants, are quickly lost from the system. When the plants are removed, the water cycle is affected, because much of the ecosystem's water was held by the plants. The local climate may become drier, with streams disappearing. Organisms that relied on a humid, shady forest disappear as well.

After an area of rainforest has been cleared, there is not much left to keep farming going beyond a few years or to provide the conditions for regrowth of the rainforest. Yet, a range of human demands, including those for wood, crops, pastures, mines, roads, and dams, continue to cause rainforest destruction.

▶ CAN WE SUCCEED IN KEEPING SOME OF
THESE FORESTS FOR THE FUTURE?

PHOTO: Shea Hazarian/creativecommons.org

There is another cost to rainforest destruction. Tropical rainforests store massive amounts of carbon in plant bodies, and this is released into the atmosphere as carbon dioxide when the forests are burned. The Amazon rainforest is estimated to store 10 billion tons of carbon in dead limbs and branches alone! This is more carbon than is contained in the amount of fossil fuels that humans burn in an entire year. So when we burn rainforests, we contribute to the buildup of carbon dioxide in the atmosphere. At the same time, we remove an important carbon dioxide

"sink," or resource that pulls carbon dioxide back out of the atmosphere.

What can we do about the pressures we are putting on these valuable ecosystems? International attention to tropical rainforest destruction has altered the way many people do business. Conservation organizations have advocated for practices to protect rainforests. For example, some industries have committed to avoid using products that destroy rainforests. Timberland, which makes leather boots, agreed to stop using leather that comes from any areas of the Amazon that were newly deforested. Cadbury Chocolates in New Zealand agreed to stop adding palm oil to its milk chocolates because palm oil production was destroying Asian rainforests. Some rainforest species that were heading for extinction, like the Scott's Tree Kangaroo, have been recovered thanks to conservation groups helping communities that live near rainforests change their destructive practices. Ecotourism has become an income-generating activity for many rainforest communities that previously relied on slash-and-burn agriculture. ∎

DISCUSSION QUESTIONS

1. Why have tropical rainforests not provided high-quality farmland, despite the millions of acres that have been cleared?

2. What actions could you or your community take that might make a difference for tropical rainforests?

Glossary

abdomen: A segment of the body of many animals. The abdomen is the third body segment of insects. *See also* thorax.

abiotic: Having to do with nonliving things.

adaptation (genetic): Any change in the structure of an organism that affects its ability to survive and reproduce in a particular environment and that may be passed to the organism's offspring through its genes.

antennae: Appendages of an insect's head used for smelling and touching.

anterior: Toward the front, or head, of an animal body.

asexual reproduction: The process by which new organisms are formed from a single parent without the union of male and female sex cells. The new organisms are genetically identical to the parent.

bacteria: Tiny unicellular organisms that lack a defined nucleus.

biology: The study of things that are or were once living.

biotic: Having to do with living things.

budding: A form of asexual reproduction during which an outgrowth of an organism, formed through cell division, breaks off and becomes an independent organism.

carnivore: Any flesh-eating organism.

castings: Solid wastes released by earthworms.

cell: The smallest organized unit of living protoplasm.

cell membrane: The outermost living layer of a plant or animal cell that regulates what enters and leaves the cell.

cell wall: The outermost, rigid, nonliving layer of a plant cell.

chlorophyll: A green plant pigment that traps energy from the sun.

chloroplast: A chlorophyll-containing plastid in plant leaf and stem cells.

chromosome: A body composed of DNA, which can be seen in a nucleus during mitosis. *See also* deoxyribonucleic acid (DNA).

cilia (singular: cilium)**:** Tiny, hair-like extensions of cells that aid in movement.

clitellum: A saddle-like structure that surrounds part of an earthworm and produces mucus, which forms a sheath around mating earthworms and a cocoon around their eggs.

community: All of the living things in an area.

compound light microscope: A microscope that uses two lenses and light to make a specimen visible.

coverslip: A piece of glass or clear plastic that is placed over the specimen on a microscope slide.

cyst: A tough, protective envelope that forms around certain microorganisms.

cytoplasm: The living material within the cell membrane.

decomposition: The breaking down of a substance into simpler substances.

deoxyribonucleic acid (DNA): Hereditary material of which chromosomes are comprised. *See also* chromosome.

depression slide: A microscope slide that has a concave area in which to put a drop of liquid or a specimen.

digestion: The breakdown of food into simpler particles that can be used as nutrients by an organism.

dormancy: A period during which organisms reduce their level of activity to a minimum to survive unfavorable environmental conditions.

dry-mount slide: A microscope slide on which no water is used.

ecosystem: A community of organisms interacting with their abiotic environment.

endangered species: A species that is at risk of becoming extinct in the near future.

evolution: A theory that states that organisms have descended from earlier forms and involves changes in organisms' genetic makeup, which are passed on through many generations.

excretion: The process by which animals eliminate waste products.

exoskeleton: A hard outer shell that covers the bodies of certain animals, including crustaceans and insects.

fermentation: A type of cellular respiration that occurs in plants and some yeasts and does not require oxygen.

field of view: The maximum area that is visible through the lens of a microscope.

flagellum (plural: flagella)**:** Whip-like extension of a cell that aids in movement.

focus: To adjust the position of a lens in order to make a clear image.

food chain: A linear pathway showing feeding relationships between organisms in an ecosystem.

food web: A weblike diagram that shows the feeding relationships among organisms in an ecosystem.

fragmentation: A form of asexual reproduction in which a piece of an organism breaks off and regenerates into a new organism.

frond: The leaves of ferns and certain aquatic plants, such as *Lemna*.

gene: One of many portions of a DNA molecule that contains genetic instructions.

genus (plural: genera)**:** A category of biological classification that ranks between family and species; always the first part of a scientific name; written in Latin or Greek.

habitat: The place where an organism naturally lives.

herbivore: Any plant-eating organism.

hyphae: The body of a typical fungus, consisting of many tiny tubes.

invertebrate: An animal without a backbone. *See also* vertebrate.

lens: A piece of curved glass or other clear material that bends light rays. Lenses can help make things look clearer, larger, and closer.

life cycle: The stages an organism goes through from conception through death.

life process: One of many processes, such as respiration, digestion, or reproduction, required for an organism to survive.

microorganism: An organism that cannot be seen without magnification.

mycelium: The vegetative body of a fungus.

natural selection: The process by which organisms that are better able to deal with changes in their environment tend to survive and reproduce, passing their desirable traits on to their offspring.

nematocyst: Poisonous thread in the sting cells of certain organisms, such as *Hydra* and jellyfish.

nucleus: The "command center" of the cell; regulates cell functions and contains the DNA.

objective lenses: Lenses of different magnifications on a microscope.

offspring: A new organism that results from asexual or sexual reproduction.

omnivore: An organism that eats both flesh and plant matter. *See also* carnivore; herbivore.

organ: A group of tissues working together to perform a specific function.

organelle: One of many structures in a cell that performs a specific function.

organism: A living creature.

parasitic: Obtaining nutrition from living on or in another organism.

photoreceptor: A structure or pigment sensitive to light.

photosynthesis: The process by which chlorophyll-containing cells use energy from the sun to combine water and carbon dioxide to produce glucose and to release oxygen as a byproduct.

plastid: Plant cell organelles containing pigments.

population: A group of individuals of a species occupying a specific region.

posterior: Toward the back end of an animal body.

protist: A group of one-celled organisms with well-defined nuclei that are not classified as plants, animals, or fungi.

protoplasm: A general term for the living material within a cell.

pseudopod: The "false foot" of protists such as an amoeba, which is composed of flowing cytoplasm. In amoebae, they aid in movement and in capturing food.

regeneration: The process by which certain organisms produce new body parts.

reproduction: The process of creating organisms of the same species.

scanning electron microscope (SEM): An instrument that bounces electrons off objects to create a three-dimensional image that is more highly magnified than possible through a light microscope.

scientific name: A universally used name for an organism; consists of two words representing the organism's genus and species. Scientific names are derived from Latin or Greek terms.

segment: A body section of an organism.

setae: Tiny, hair-like structures on the body of certain annelids, such as earthworms, that help them grip a surface.

sexual reproduction: Reproduction that is accomplished through the union of a male sperm and a female egg.

species: The last part of a scientific name, ranked after genus in biological classification; also applies to a group of interbreeding organisms that share similar characteristics.

spore: A sexual or asexual reproductive cell of an organism.

stamen: The male reproductive organ of flowering plants.

succession: A series of progressive changes in the plant and animal life in an area that can lead to a stable community in which the numbers and types of organisms are in relative balance.

symbiotic: A term that describes a relationship between two organisms in which both organisms benefit.

taxonomy: The science of classifying living things.

thorax: In insects, the body part between the head and abdomen. *See also* abdomen.

threatened species: A species that is at risk of becoming endangered in the near future.

tissue: A group of cells working together to perform a specific function.

trait: An inherited characteristic of an organism.

variable: A factor in an experiment that can be changed and measured.

Venn diagram: A diagram that shows similarities and differences within and among groups of different things.

vertebrate: An animal with a backbone. *See also* invertebrate.

wet-mount slide: Two microscope slides, or a slide and coverslip, with a drop of liquid and/or a specimen between them.

Index

Photo Credits

Front Cover
John and Karen Hollingsworth/U.S. Fish and Wildlife Service

4 Eric Long, Smithsonian Institution
10 Courtesy of the National Library of Medicine **11** Charlotte Raymond, Photographer (top right) Justin B. Boyles (bottom right) Mark Bray/creativecommons.org **14** (top) Tijl Vercaemer/creativecommons.org (bottom) B.M. Drees, Texas AgriLife Extension Service, Texas A & M University **16** Philp Kahn/©2010 University of California/creativecommons.org **17** Bruce Avera Hunter/life.nbii.gov **18** Steve Hillebrand/U.S. Fish and Wildlife Service **19** Peggy Greb, Agricultural Research Service/U.S. Department of Agriculture **20** Courtesy of Henry Milne/© NSRC **22** Jodiepedia/creativecommons.org **32** Chip Clark, National Museum of Natural History, Smithsonian Institution **33** (top) Michel Lecoq (CIRAD) (bottom) Jack Dykinga, Agricultural Research Service/U.S. Department of Agriculture **35** (left) Courtesy of Carolina Biological Supply Company (right) Courtesy of R.W. Matthews **38** (top) Library of Congress, Prints & Photographs Division, LC-USZ62-95187 (bottom) Library of Congress, Prints & Photographs Division, LC-USZ62-110443 **39** Photo by Eric Erbe, digital colorization by Chris Pooley, Agricultural Research Service/U.S. Department of Agriculture **40** National Science Resources Center **43** (left) Courtesy of Henry Milne/© NSRC (right) Courtesy of Henry Milne/© NSRC **46** Courtesy of Henry Milne/© NSRC **50** (top left) William R. West/Carolina Biological

Supply Company (top right) William R. West/Carolina Biological Supply Company (bottom) Courtesy of Carolina Biological Supply Company **51** Courtesy of Carolina Biological Supply Company **52** Courtesy of Carolina Biological Supply Company **53** Ed Bierman/creativecommons.org **54** Robbie Sproule/creativecommons.org **56** Pete Pattavina/U.S. Fish and Wildlife Service **61** Courtesy of Henry Milne/© NSRC **62** (left) Courtesy of Carolina Biological Supply Company (right) Andrew Ross/creativecommons.org **63** NOAA's National Weather Service (NWS) Collection **64** Courtesy of Carolina Biological Supply Company **66** Dartmouth Electron Microscope Facility, Dartmouth College **67** Courtesy of Carolina Biological Supply Company **73** Courtesy of Carolina Biological Supply Company **74** (top) Courtesy of Carolina Biological Supply Company (bottom) Courtesy of Carolina Biological Supply Company **76** chadh/creativecommons.org **77** (top left) foshie/creativecommons.org (right) James Gathany, CDC/Dr. Christopher Paddock (bottom) Frank Kovalchek/creativecommons.org **78** Axel Rouvin/creativecommons.org **79** USDA Soil Biology Primer **80** Charlotte Raymond, Photographer **85** Courtesy of Henry Milne/© NSRC **87** Courtesy of Henry Milne/© NSRC **91** (left) Courtesy of Carolina Biological Supply Company (right) Courtesy of Carolina Biological Supply Company **92** Miloslav Kalab **93** Courtesy of Goucher College **95** Scott Bauer, USDA Agricultural Research Service, Bugwood.org **96** (top) Courtesy of Carolina Biological Supply Company (bottom) Courtesy of Carolina Biological Supply Company **98** Ryan Hagerty/U.S.

Fish and Wildlife Service **100** (left) Courtesy of Carolina Biological Supply Company (right) Courtesy of Carolina Biological Supply Company **101** (top left) Courtesy of Carolina Biological Supply Company (top right) Courtesy of Carolina Biological Supply Company (bottom) Courtesy of Carolina Biological Supply Company **103** National Science Resources Center **104** Courtesy of the National Library of Medicine **106** National Science Resources Center **107** (top) (2205) Are We Underestimating Species Extinction Risk? PLos Biol 3(7): e253. doi: 10.1371/journal.pbio.0030253/Creative Commons Attribution 2.5 License (bottom) NASA Ames Research Center **109** Jim Kuhn/creativecommons.org **110** Courtesy of Carolina Biological Supply Company **111** Courtesy of Carolina Biological Supply Company **112** emmett anderson/creativecommons.org **114** National Science Resources Center **116** John J. Mosesso/life.nbii.gov **117** Courtesy of Carolina Biological Supply Company **118** (left) Courtesy of Carolina Biological Supply Company (right) Courtesy of Carolina Biological Supply Company **120** (top) Jose Hernandez @ USDA-NRCS PLANTS Database (bottom) Christian Fischer/creativecommons.org **121** National Science Resources Center **124** Paul Hebert/Gewin V (2005) Functional Genomics Thickens the Biological Plot. PLoS Biol 3(6): e219. doi:10.1371/journal.pbio.003029/Creative Commons Attribution 2.5 License **126** (top) Courtesy of Carolina Biological Supply Company (bottom) Courtesy of Carolina Biological Supply Company **128** Courtesy of Carolina Biological Supply Company **134** Peter J. Bryant, University of California, Irvine **143** Courtesy of Carolina Biological Supply Company **144** U.S. Marine Corps **145** Andrew Braithwaite/creativecommons.org **147** Mr. Mohammed Al Momany, Aqaba, Jordan/NOAA's Coral Kingdom Collection **148** Tim Parkinson (flickr.com/timparkinson/)/Creative Commons Attribution License (http://creativecommons.org/licenses/by/2.0) **150** gr33n3gg/creativecommons.org **152** Courtesy of Carolina Biological Supply Company **153** (top) Courtesy of Carolina Biological Supply Company (bottom) Courtesy of Carolina Biological Supply Company **155** (top left) polandeze/creativecommons.org (bottom left) Courtesy of Carolina Biological Supply Company (right) Anssi Koskinen/creativecommons.org **157** (top) Sandy Richard (bottom) Keith Weller, Agricultural Research Service/U.S. Department of Agriculture **158** César Astudillo/creativecommons.org **161** (top) Library of Congress, Prints & Photographs Division, LC-USZ62-3499 (bottom) ©2010 Carolina Biological Supply Company **163** by T.K. Maugel, Laboratory for Biological Ultrastructure, University of Maryland, College Park **164** (top) RonAlmog/creativecommons.org (right) Scott Bauer, Agricultural Research Service/U.S. Department of Agriculture **165** Library of Congress, Prints & Photographs Division, LC-USZ62-93513 **166** David Burdick, NOAA's Coral Kingdom Collection **168** (right) John J. Mosesso/life.nbii.gov (left) J.E. Appleby/U.S. Fish and Wildlife Service **173** (top) vaka0627/creativecommons.org (left) Smithsonian Institution, Carl C. Hansen, #92-4630 **174** Shea Hazarian/creativecommons.org **175** (top) Alexander Torrenegra/creativecommons.org (bottom left) psyberartist/creativecommons.org (middle) Jessie Cohen, Smithsonian National Zoo (right) Marcos Guerra/Smithsonian Tropical Research Institute (bottom right) Alexander Torrenegra/creativecommons.org **176** mattmangum/creativecommons.org **177** Shea Hazarian/creativecommons.org